"Taking to heart critiques of progressive and salvific "universal" history, and of linear, homogeneous, conceptions of time consonant with it, this book gracefully sets aside any nostalgia for such time and history, and any desire to redeem an idea of the political still tethered to a wishful consensual time. Rahman gently, sincerely, and fluently compels us to think difference in terms of dissonant but coexisting regimes of time, to moor ourselves ethically to the unruly and undisciplined nature of time as well as memory, and to continue to think, imagine, and desire global justice from these necessary dislocations. The book is refreshingly free of the defensive and dizzying anxieties that infect many other proposed politics of contingency, as it is interested not in lamenting our fragilities or bemoaning the curses of history and time, but in what we might be keeping from happening when we do so, and in what our expectations for and from political action might look like if we embraced contingency as a constitutive and affirmative principle of politics rather than a problem for it. In doing so, this book performs its message. It is, thus, a gift for any reader interested in articulating a shared possibility in a discrepant world".

—Asma Abbas, Bard College at Simon's Rock

"*Time, Memory, and the Politics of Contingency* at once argues for and enacts a prismatic image of time by excavating disjointed experiences of temporality past and present, from both within and beyond the familiar coordinates of the Western philosophical tradition. An eloquent call to take the disobedient rhythms of lived time not as problem in need of regularization and discipline, but as a clue to the eddies of memory and anticipation that shape and constrain the possibilities of political action in the present".

—Roxanne L. Euben, Ralph Emerson and Alice Freeman Palmer Professor of Political Science, Wellesley College

Time, Memory, and the Politics of Contingency

In recent years, there has been an increased attention to temporality in political theory, and such attention is sorely needed. For too long, political theory, with the exception of occasional phenomenological forays, has remained grounded in a particular experience of time as linear and sequential. This book aims to unsettle the dominant framework by putting time itself, and the experience of time in everyday life, at the center of its critical analysis.

Smita Rahman focuses on the experience of time as one where past, present, and future intermingle with each other and refuse to adhere to a sequential structure. Rather than trying to tame the flux of time, this book places this "out of joint" experience of time at the center of its analysis of global politics. Rahman takes the highly abstract concept of time and de-centers it to speak to a wide range of political issues across disciplines. She does so by exposing the cultural construction of the foundational concept of time in political theory and attending closely to the challenges of cultural incommensurability that it encounters in a globalized world of difference. Specifically, the book looks at interrogation practices in Afghanistan; the challenges of coping with the burdens of collective memory in Algeria, South Africa, and Rwanda; the difficulty of uncritically applying such a framework to the Muslim world through the language of secularism; and finally at the beginnings of democratic emergence in Bangladesh to explore a politics of contingency.

By focusing on issues of contemporary global politics through the lens of political theory, this book draws on literature across disciplines and explores the complex image of time by engaging the work of thinkers for whom time and memory have emerged as a critical issue of analysis, and unpacking the politics of contingency that emerge from such a reading. The book's new insights on political temporality will interest scholars of contemporary political theory, comparative political theory, critical theory, human rights, conflict studies, and religion and politics.

Smita A. Rahman is Associate Professor of Political Science at DePauw University, where she teaches courses in contemporary and comparative political theory, focusing on the politics of memory and on secularism and its tensions. Her current project is an examination of the politics of honor in literature and visual culture.

Routledge Innovations in Political Theory

For a full list of titles in this series, please visit www.routledge.com

26 Logics of Critical Explanation in Social and Political Theory
Jason Glynos and David Howarth

27 Political Constructivism
Peri Roberts

28 The New Politics of Masculinity
Men, Power and Resistance
Fidelma Ashe

29 Citizens and the State
Attitudes in Western Europe and East and Southeast Asia
Takashi Inoguchi and Jean Blondel

30 Political Language and Metaphor
Interpreting *and* Changing the World
Edited by Terrell Carver and Jernej Pikalo

31 Political Pluralism and the State
Beyond Sovereignty
Marcel Wissenburg

32 Political Evil in a Global Age
Hannah Arendt and International Theory
Patrick Hayden

33 Gramsci and Global Politics
Hegemony and Resistance
Mark McNally and John Schwarzmantel

34 Democracy and Pluralism
The Political Thought of William E. Connolly
Edited by Alan Finlayson

35 Multiculturalism and Moral Conflict
Edited by Maria Dimova-Cookson and Peter Stirk

36 John Stuart Mill—Thought and Influence
The Saint of Rationalism
Edited by Georgios Varouxakis and Paul Kelly

37 Rethinking Gramsci
Edited by Marcus E. Green

38 Autonomy and Identity
The Politics of Who We Are
Ros Hague

39 Dialectics and Contemporary Politics
Critique and Transformation from Hegel through Post-Marxism
John Grant

40 Liberal Democracy as the End of History
Fukuyama and Postmodern Challenges
Chris Hughes

41 Deleuze and World Politics
Alter-globalizations and Nomad Science
Peter Lenco

42 Utopian Politics
Citizenship and Practice
Rhiannon Firth

43 Kant and International Relations Theory
Cosmopolitan Community Building
Dora Ion

44 Ethnic Diversity and the Nation State
National Cultural Autonomy Revisited
David J. Smith and John Hiden

45 Tensions of Modernity
Las Casas and His Legacy in the French Enlightenment
Daniel R. Brunstetter

46 Honor
A Phenomenology
Robert L. Oprisko

47 Critical Theory and Democracy
Essays in Honour of Andrew Arato
Edited by Enrique Peruzzotti and Martin Plot

48 Sophocles and the Politics of Tragedy
Cities and Transcendence
Jonathan N. Badger

49 Isaiah Berlin and the Politics of Freedom
"Two Concepts of Liberty" 50 Years Later
Edited by Bruce Baum and Robert Nichols

50 Popular Sovereignty in the West
Polities, Contention, and Ideas
Geneviève Nootens

51 Pliny's Defense of Empire
Thomas R. Laehn

52 Class, States and International Relations
A Critical Appraisal of Robert Cox and Neo-Gramscian Theory
Adrian Budd

53 Civil Disobedience and Deliberative Democracy
William Smith

54 Untangling Heroism
Classical Philosophy and the Concept of the Hero
Ari Kohen

55 Rethinking the Politics of Absurdity
Albert Camus, Postmodernity, and the Survival of Innocence
Matthew H. Bowker

56 Kantian Theory and Human Rights
Edited by Reidar Maliks and Andreas Follesdal

57 The Political Philosophy of Judith Butler
Birgit Schippers

58 Hegel and the Metaphysical Frontiers of Political Theory
Eric Lee Goodfield

59 Time, Memory, and the Politics of Contingency
Smita A. Rahman

Time, Memory, and the Politics of Contingency

Smita A. Rahman

Routledge
Taylor & Francis Group
NEW YORK AND LONDON

First published 2015
by Routledge
711 Third Avenue, New York, NY 10017

and by Routledge
2 Park Square, Milton Park, Abingdon, Oxon OX14 4RN

Routledge is an imprint of the Taylor & Francis Group, an informa business

First issued in paperback 2016

Library of Congress Cataloging-in-Publication Data

Rahman, Smita A.
Time, memory, and the politics of contingency / by Smita A. Rahman.
pages cm. — (Routledge innovations in political theory ; 59)
1. Political science—Philosophy. 2. Time—Political aspects.
3. Chronology, Historical—Political aspects. 4. Collective memory—Political aspects. I. Title.
JA71.R23 2014
320.01—dc23
2014016003

ISBN13: 978-1-138-78272-3 (hbk)
ISBN13: 978-1-138-68701-1 (pbk)

Typeset in Sabon
by Apex CoVantage, LLC

For Rob

Contents

Acknowledgments

This book would never have come into being without the generosity of several people. First and foremost, the project was nurtured in the uniquely supportive community of political theory graduate students at Johns Hopkins University. Simon Glezos, who has always shared my theoretical sensibility in a deep and meaningful way, was central to the growth and development of this project, from its earliest conceptualization in graduate school to his close and thoughtful reading of the book manuscript in recent months. Matthew Scherer, Lars Tønder, and Paulina Ochoa pushed me critically from every angle and nudged the project in productive directions and steered me away from my own errors. I remain grateful for their deep intellect, continued support, and the warmth of their friendship. Outside the seminar room, Erin Ackerman and Blake Ethridge provided me with the kind of deep and genuine friendship that made writing this book possible by sprinkling my writing days with humor and much needed distraction.

Some of us are blessed and fortunate to have a mentor who goes above and beyond any designated role to genuinely infuse your being with their deep intellect and sheer generosity. Bill Connolly is such a person. I cannot thank him enough for his vigorous and detailed critiques, for his unconditional support in both work and life, for his continued generosity and deep friendship, and for constantly pushing me to think beyond what seemed to be my limits. I would never have been nudged along so many lines of flight without him. At Hopkins, Dick Flathman took me under his wing from my very first day, continually prodded me with his sharp and unrelenting questions, and constantly made me smile with his dry wit. Jane Bennett pushed me to be a better writer and a more fluid thinker with her deeply intelligent and imaginative readings of texts and inspired me with the sheer joy of her being.

I came to political theory by chance, through a brief introduction to the canonical texts in an introductory political science class taught by the brilliant Ian Lustick in the first semester of my freshman year at the University of Pennsylvania. The very next semester I signed up to take my first theory class with Anne Norton and it set me on my path. Anne dazzled me with her intellect, her deep reverence for texts, her surprising humor, and her always evident appetite for learning. She encouraged me to take both undergraduate and graduate theory courses at Penn, introduced me to my first set of readings on the politics of time and memory, which made a lasting impact on me, and pushed me to do serious undergraduate research in my senior thesis that became the seed from which this book would grow. I have never stopped wanting to grow up and be Anne Norton. She was and continues to be a source of tremendous inspiration.

I have been very fortunate to have been nurtured by a larger community of exceedingly generous political theorists as well. Elisabeth Anker has been tremendously helpful with her insightful comments on my book proposal and her continued encouragement of the book project. I am so grateful for her friendship. Roxanne Euben gave me detailed, incisive comments on a portion of the manuscript and continues to provide support and constant inspiration. Shirin Deylami, Andrew March, Asma Abbas, Mario Feit, Kennan Ferguson, Steve Johnston, Nikolas Kompridis, and Samuel Chambers all provided support and encouragement at key stages of the development of this project as well. I thank them all for the rich and illuminating conversations and for their remarkable kindness. I also want to thank my fellow panelists at several conference meetings, particularly those at the Western Political Science Association and American Political Science Association for their engagement with several chapters of this book in its draft stages. Their comments and questions were very helpful in improving and clarifying the central arguments of this book.

At Routledge, Natalja Mortensen has been a wonderful editor, who seized on the project with enthusiasm from the very beginning and provided timely feedback and encouragement throughout the process. Two anonymous reviewers for Routledge provided useful criticism and allowed me to see the broader contributions of the project more clearly. I also want to acknowledge the journal *Contemporary Political Theory* for allowing me to use my article for this publication, an extended version of which forms Chapter 4 of this book and the *Journal of Islamic Law and Culture* for the use of some key arguments from my article in sections of Chapter 5.

I am fortunate to work at an institution that embodies the ideal of a liberal arts college, with its dedication to teaching at the highest level and its commitment to faculty research. DePauw University has been a source of tremendous support without which I could not have completed this book. I thank them for the multiple Fisher course reassignments and summer stipends, the essential pre-tenure leave, and the additional faculty development and institutional funds they have provided to support this project. In the Political Science Department, I am surrounded by generosity and deep personal warmth. Clarissa Peterson has provided essential support from my first day on the job and Brett O'Bannon has been a consistent model of generosity and intellectual curiosity. Sunil Sahu has been a tremendously supportive chair and mentor and provided important guidance throughout my five years at DePauw, as has Bruce Stinebrickner with his deep experience.

Deepa Prakash and Maryann Gallagher have proved time and again that it is possible to make great friends and have way too much fun at work. I thank them for their continued friendship and support and the constant laughs.

There is a vibrant community of young researchers at DePauw who continue to do critical and cutting-edge work in their fields. Mona Bhan, David Alvarez, and Nahyan Fancy embody this ideal and have been very useful in providing feedback as part of a research writing group. I have had too many wonderful students at DePauw to count over the last few years, and I thank them all for their curiosity and energy, particularly those who have taken my senior seminar on the politics of memory and critically engaged with some of the arguments of this book with passion and excitement. In particular, I want to thank Stephen Worden, David Tykvart, Anajah Roberts, Stephanie Grauer, and Patrick Tatsuya Schmitz, for providing me with the rarest and most meaningful of exchanges in pushing me to expand my own research horizons with their exceedingly thoughtful and subtle questions in the classroom. They continue to inspire me every day, even as they move on in their own lives and careers.

The past few years have been marked by a deep and indelible personal tragedy that emerging from proved very difficult. I want to thank a group of wonderful and supportive women who I met during my darkest moments and who are now my friends for life. They taught me how to suffer and endure and to find the courage to reach for joy again.

I am blessed with a remarkable family that is an immense source of affection and grounds me in everything I do. My deepest thanks go to my Ammu and Baba for their unconditional love and support, for teaching me how to write and think, for giving me the gift of travel and the more important one of imagination, for teaching me that no barrier is insurmountable, and for their continued patience with me. My sister, Ushshi, remains a source of continued inspiration with her brilliant writing. I have always been and continue to be grateful to my Nanu for being such a constant and dedicated presence in my life, without whom very little would be possible.

Finally, to my two guys, who are my everything. To my Kai, who is my deepest source of joy and laughter, who makes the most mundane of days magical with his fearlessness, his empathy for others, his boundless curiosity, and the vividness of his inexhaustible imagination. He is truly the "and" in the most meaningful sentence in my life. To Rob Watkins, for reading every inch of this manuscript multiple times and providing the most detailed set of comments at every turn. This book would not be possible in any form without you, from its earliest inspiration to its completion. But most importantly, for being my *jaan*, my partner in life and love, for the deepest of friendships, the constant patience and reassurance, the forever glowing warmth of shared recognition, and the continually inspiring conversations. I love and treasure the life we have built together.

Introduction

Time, in the familiar rhythms of lived experience, often seems to be something other than the carefully calibrated and quantified progression of hours, minutes, or seconds marked by clock-time. Rather, the experience of time is that of a fluid mingling of the tenses, where the present and the past coexist with each other and gnaw into the future, where each tense contains the impress of the other. Consider, for instance, sitting at a café in a cozy armchair and how the feeling is amplified by the memories of previous afternoons spent in that same spot. In that moment, the layers of past memories mingle and coexist with the present, lending it depth and vibrancy. Or consider the earthy smell of rain as it perfumes the air and how that affective reminder sends you searching your past for the memory of a monsoon evening years ago. Or perhaps the visceral tension you feel knotting in your stomach as you anxiously wait for an interview, when the pressure of the future colliding into the present is palpable. This is a complex experience of time—where time does not seem neatly sequential, or obediently linear, where the past, present, and future do not seem to follow each other in calibrated intervals.

In such moments of everyday occurrence, time is often experienced in fluid and complex ways—as a stormy sea in which the waves of the past and the present crash against those of the future in a manner that is both rhythmic and dissonant, both transient and eternal. Paying attention to this temporal experience often entails the recognition that time is "out of joint" as Shakespeare's Hamlet put it. Hamlet's diagnosis has become shorthand in recent years for referring to the experience of time as something other than clock-time. The use of the reference is heavily indebted to Derrida's magisterial analysis of the appearance and reappearance of the ghost of Hamlet's father in *Specters of Marx*. In Derrida's words, "We are questioning in this instant, we are asking ourselves about this instant that is not docile to time, at least to what we call time. Furtive and untimely, the apparition of the specter does not belong to that time, it does not give time, not that one".[1] This experience of time that is not docile and which refuses to adhere to the sequential structures that try to contain it is at the heart of this book. As Derrida puts it, it is a "disjointed or disadjusted now, "out of joint", a disjointed now that always risks maintaining nothing together in the assured conjunction of some context whose border would still be determinable".[2]

What then is this "disjointed" and "disadjusted" moment that coheres only to slip away and seemingly escape any attempt to define it? In Shakespeare's telling, Hamlet is beckoned by his father's ghost, is burdened with the knowledge of his murder and betrayal at the hands of his uncle, and is tasked with the act of revenge. In response to this haunting past flooding his present, Hamlet famously exclaims, "The time is out of joint: O cursed spite, / That ever I was born to set it right!"[3] In this proclamation, Hamlet, the quintessential modern protagonist, speaks of the vertiginous experience of time. He is reeling from loss, grief, and betrayal and is consumed by the heavy burden of avenging his father, who imprints on him a ghostly demand from his past and simultaneously wrenches his future into the present. "Out of joint", for Hamlet, signifies a world without moorings, where time has been unclasped from its sequential structure and is a swirling eddy of confusion. His father's ghost is literally the spectral presence of his past—a haunting reminder who arrives to depict his uncle as a murdering usurper and his mother as an adulterous accomplice. The tenses are collapsed together for Hamlet in dizzying form and must be put back in order through his actions. His tragic burden is to somehow overcome this vertigo and to negotiate a path through its midst, in a context whose borders can never be fully determined.

What lessons might Hamlet's diagnosis offer for politics in a world where time is experienced in the complex confluence of the tenses? In his fleeting but profound comments on Hamlet from *The Birth of Tragedy*, Nietzsche provides us with some tantalizing clues. He writes that

> In this sense the Dionysian man resembles Hamlet: both have once looked truly into the essence of things, they have *gained knowledge,* and nausea inhibits action; for their action could not change anything in the eternal nature of things; they feel it to be ridiculous or humiliating that they should be asked to set right a world that is out of joint. Knowledge kills action; action requires the veils of illusion: that is the doctrine of Hamlet, not that cheap wisdom of Jack the Dreamer who reflects too much and, as it were, from an excess of possibilities does not get around to action. Not reflection, no—true knowledge, an insight into the horrible truth outweighs any motive for action, both in Hamlet and in the Dionysian man.[4]

The understanding of time as "out of joint" is "true knowledge" for Nietzsche; indeed, it is a fundamental insight into the condition of modern life. It is not mere "reflection" but rather a deep recognition that arrives with a heavy burden. In this sense, Nietzsche's tragic understanding of time as "out of joint" runs counter, in some respects, to another poetic and romanticized account in modern literature—Marcel Proust's deeply affective account of the madeleine and the evocative past that swirls up with every lemon-scented bite. Instead of bathing oneself in a flavorful and pleasant past, for Nietzsche, Hamlet's knowledge leads to nausea and "kills action".[5] This is a deeply tragic insight that underlines what might at first glance appear to be the antipolitical consequences of such knowledge. If understanding a past that never fully disappears but insistently forces its way into the present paralyzes action, it could erode all possibility for political action. As Nietzsche states, this is when "the danger to his will is greatest".[6] However, a world in which we recognize that time is "out of joint", where we know the tenses are disobedient and unruly and will not stay neatly demarcated in place, is not necessarily a world without the possibility of political action. Nietzsche continues, "Here, when the danger to his will is greatest, *art* approaches as a saving sorceress, expert at healing. She alone knows how to turn these nauseous thoughts about the horror or absurdity of existence into notions with which one can live".[7] There is hope in the actions of the creative will, which can tame the nausea and transform it into something affirmative.

What then are the political implications of this recognition of "a world that is out of joint"? How do we experience this recognition and what might be the value of exploring contemporary politics from this temporal perspective? In our everyday lives we experience out of joint-ness in multiple ways. It can be a fragment of the past that hurtles into our present in a jarring and dissonant form, such as with the memory of trauma from abuse, which makes the present a muddle of suffering and dislocation. Or the past can make its presence known in loving and warm hues when you look at your child and remember his first steps. On a collective level, out-of-joint politics can be one of nostalgia and idealized remembrance, such as when we call up the revolutionary past of American founding to imagine a community founded on American Exceptionalism). Or it can be one where whole communities reckon with the consequences of genocide by their neighbors, where the past refuses to withdraw its vicious claws from the present, where it kills action, where the only choices seem to be whether to forget or to remember.

Such a mode of politics operates in a world without the security of foundations. Not only does it require attending to the disruption of time, but it also requires coming to terms with a globalized world of difference where multiple culturally constituted images of time jostle and disjoint each other. It operates in a world of violence, misunderstanding, and incommensurability that can paralyze action. But it is also suggests that a more complex form of politics is possible, one that embraces contingency and the fluidity of the tenses at its core and seeks affirmative, if necessarily partial, solutions to problems. It calls for a politics that tries to appreciate and understand difference, which eschews certainty and stability to ride the waves of contingency.

This book seeks to explore this complex image of time and the politics that unfolds from it by engaging the work of thinkers for whom time and memory have emerged as a central issue of analysis. First, it argues that it is essential to pay attention to the complexity and multiplicity of temporal frameworks in a world replete with cultural difference. Second, it argues that the dominant concept of time that structures much of political theory conceals in its midst an appeal to a transcendent being or moral law to impose order and structure without which humans are abandoned to a vertiginous state of confusion, like the nausea invoked by Nietzsche. Third, it seeks to excavate a complex image of time that captures the flux and flow of time in lived experience, where time is recognized as being out of joint. This is an image of time where the neat sequential categories are dislodged and the past, present, and future rub up against each other in the moment, mingling and combining with each other and giving each moment its distinctive hue. Fourth, it asks what the political implications of such a complex image of time might be. If the past is inescapably present, how do societies that live with grievously injured pasts come to terms with them in the present for the sake of the future? Fifth, it critically examines to what extent it is possible to continue to speak of secular time today—a homogeneous experience of time in which we can universally participate—and explores the consequences of deploying such a framework in the contemporary discourse of the West's relationship with Islam. Finally, it concludes with an exploration of the fragility of democratic emergence and the tragic possibility that it represents when the present moment always remains contingent on the disruptiveness of time.

Drawing attention to the complexity and multiplicity of temporal frameworks is unavoidable in an interconnected world marked by difference, where various cultural constituencies experience time differently. There are familiar symbols of such difference, such as calendars that mark the beginning of time on different dates based on religious significance, but there are also different speeds in which time is experienced and different ways in which we understand ourselves through commemoration and anticipation that involve temporal negotiation. The first chapter explores what the political theorist Sheldon Wolin called the "splintering" or fragmentation of a common temporal framework that could once have been simply called "our time". While we might mark our clocks in fixed deviations from Greenwich Meridian Time, we experience time in different ways. This chapter follows Wolin's analysis of the fragmentation of the common narrative of "our time" in the face of cultural difference in order to explore the consequences of this rupture. To that end, it examines the interrogation of an Afghani prisoner by Central Intelligence Agency officials as an example of an encounter between diverse zones of time which are incommensurable with each other. Such moments of out of joint-ness can be attended to, or they can be silenced by acts that ignore different experiences of temporality and insist on unity through the differential exercise of power.

The second chapter explores the dominant concept of time in Western political theory, which Wolin referred to as "our time". It is such a familiar image of time that it is seldom contested or complicated, one where time is seen as linear and sequential and structured by the universalizing assumption of progress. This chapter looks at the Christian origins of this time-image in the writings of Augustine for whom such a concept is an expression of his faith. For Augustine, the dizzying eddies of lived time must be quieted by the transcendent stillness of an eternal God. Time takes the form of the progressive realization of redemption; it is a continuum in which the sinful man is healed by grace and through acts of discipline and confession. In such an account, time begins with Genesis and the fall of man and ends with the Day of Judgment and the prospect of eternal salvation. This image of time was stripped of much of its religious significance during the Enlightenment, when Kant refashioned it for the demands of the new age. Kant's account of time at first glance seems to be a secular rendering, replacing the language of redemption so vital to Augustine's understanding with the modern idea of progress. However, Kant retains the language of moral imperative and strengthens the demand to silence the dissonance of out of joint-ness and to quell the anxieties it arouses. By reading Augustine and Kant together, the chapter highlights the structuring demands of this time image—its opposition to nausea, to contingency, and the extent to which it is a "veil of illusion", in Nietzsche's terms, which makes political action possible but comes at the cost of true knowledge of the late modern condition.

The third chapter excavates a complex account of time which, in contrast to dominant accounts in Western thought, is attentive to the dissonances created by the lived experience of time. Such accounts are scattered throughout the Western philosophical tradition, from the writings of the Stoics to the messianic account put forth by Walter Benjamin. This chapter examines Henri Bergson's work on perception and memory in order to excavate a complex image of time where the tenses flow and mingle into each other, without being trumped or quieted by faith, either religious or rational, into stillness. But Bergson's image of time alone does not pose a satisfactory answer to the key question of what such a temporal understanding means for lived experience. To that end, Nietzsche's self-confessedly "untimely" writings are investigated. This chapter looks at Nietzsche's riddle of eternal return, it explores his diagnosis of the will's resentment of the past, and it engages his efforts to forge an ethos of affirmation in order to overcome it.

The fourth chapter of this project explores the dissonances of experience in the politics of collective memory. What are the political implications of such a complex image of time? What experiences and conditions can be engaged more creatively when viewed through this prism? If the past is inescapably present, how do societies that live with grievously injured pasts come to terms with them in the present for the sake of the future? This chapter examines the way in which three countries have recently attempted to negotiate the politics of memory—Algeria, South Africa, and Rwanda. It analyzes each attempt through the prism of a complex image of time, asking, how much should be forgotten in order to free the present and how much should be remembered? It also poses the difficult question of how such pasts can be overcome and transfigured in the present for the sake of an affirmative politics of the future.

The fifth chapter returns to a critical examination of the dominant concept of time outlined in the second chapter by analyzing it in the context of contemporary politics. It asks, "To what extent is it possible to speak of secular time, of a homogeneous experience of time in which we can universally participate? How does this image of time, which structures many of our modernizing and secularizing demands in foreign policy, actually operate in contemporary politics?" It begins by examining Charles Taylor's account of secular time as "profane" time, which takes on a contrasting hue when it is juxtaposed with the contemporary discourse of the West's relationship with Islam. In recent years, the Islamic world has been characterized as a monolithic whole, one that is anti-modern by dint of what is perceived to be a broad rejection of secularism, and one that yearns for a fossilized past and remains hopelessly anachronistic in its refusal or inability to experience and inhabit secular time. In order to examine this temporal tension, it turns to the writings of Sayyid Qutb, who provides a complex temporal perspective by arguing for a renewal of the Islamic tradition and by mounting a critique of the idea of progress that sustains secular time. By reading Qutb alongside Taylor's account of homogeneous secular time, it aims to underline the insufficiency of such an image of time for a complex and interdependent world.

Finally, the concluding chapter looks closely at Bangladesh, the country of my birth, where politics always seemed to be contingent on the ghostly past of the 1971 War on Liberation, whose legacy remains in dispute and continues to unsettle the present in many ways. It analyzes the popular democratic "Shahbagh" movement in that context—a movement that emerged in recent months as a mechanism to seek justice for war criminals, but also galvanized large segments of the population with its vitality. It argues that Shahbagh remains a tragic possibility in a country that is both haunted by the demands for vengeance and jolted by the newness of its democratic emergence. By examining this fugitive moment, it articulates a politics of contingency grounded in the instability and disruption of time and explores its political implications.

This project takes its bearings from a desire to make sense of the fluidity and dissonance of time as experienced in everyday life, an "out of joint" experience that might be intuitive to us but which remains woefully undertheorized. To do so, it complicates and de-centers the dominant image of time, which pays scant attention to the fluidity of temporal experience and often insists on suppressing it. It seeks to unearth an image of time that is complex and dissonant and argues for a politics of contingency that is attentive to moments of fragmentation and contestation in a world of difference.

NOTES

1. Jacques Derrida, *Specters of Marx: The State of the Debt, the Work of Mourning and the new International*, trans. Peggy Kamuf, with an Introduction by Bernd Magnus and Stephen Cullenberg (New York and London: Routledge, 1994), p. xix.
2. Ibid., p. 1.
3. William Shakespeare, *The Tragedy of Hamlet, Prince of Denmark*, act 1, scene 5, lines 188–189, in *The Oxford Shakespeare: The Complete Works, Second Edition,* ed. Stanley Wells and Gary Taylor (Oxford: Oxford University Press, 2005), p. 691.
4. Friedrich Nietzsche, "The Birth of Tragedy," in *The Birth of Tragedy and The Case of Wagner*, trans. Walter Kaufmann (Toronto: Vintage Books, 1967), p. 60.
5. Ibid.
6. Ibid.
7. Ibid.

1 The Limits of "Our Time"

In recent years, political theorists have become increasingly interested in the language of temporality. While the attention to examining temporality is welcome, many of these efforts seem to approach the encounter indirectly and, often, hesitantly. The small and growing attention to temporality in contemporary political theory is rather controlled, leaving the analytic structure of time essentially unchanged. There is real promise in these collective efforts—in recent books, articles, and conference papers—but there is frustration borne of them as well. The promise lies in finally turning a critical eye on one of the fundamental concepts of political theory: a metaphysical skeleton, which we have taken for granted but have rarely stopped to examine. We have yet to look at how it gives form to our body, to the crook of the elbow, to the sharp line of the jaw. We layer on to the image of time our critiques of liberalism, tolerance, freedom, and subjectivity, but we don't stop to fully critically examine the foundation that undergirds them, that which gives them their pointy angles, which jut out in sharp and uncomfortable ways.

To the extent that the recent literature on time and politics makes us aware of its bony presence, it is welcome indeed. These works reveal the metaphysical structure to be worthy of our attention, as a presence that needs to be interrogated and laid bare by the critical eye. The frustration, however, is that many of these attempts fall somewhat short of engaging the dominant image of time that undergirds most analyses of politics, which is linear, sequential, and structured by the idea of progress. Nor do they focus on uncovering a more complex image of time, which can attend to the moments of out of joint-ness we encounter in our everyday lives, where time often seems to be something other than the carefully calibrated and quantified progression of hours, minutes, or seconds marked by clock-time. Rather, the experience of time is that of a fluid mingling of the tenses, where the present and the past coexist with each other and gnaw into the future, each tense containing the impress of the other. In such moments of everyday occurrence, time is often experienced in fluid and complex ways—as a stormy sea in which the waves of the past and the present crash against those of the future in a manner that is both rhythmic and dissonant, both transient and eternal.

According to many of these recent accounts of time and politics, the language of temporality helps us to understand social formations, a world of becoming, the connection between politics and history, creative forms of evolution, and capitalism and its processes.[1] They routinely use the language of the untimely, but the concept of time itself rarely takes center stage. Samuel Chambers's *Untimely Politics*, for instance, is valuable for its attempt to underline the spectral model of approaching history through its careful and sophisticated attention to language.[2] But in the refusal to define the untimely itself, Chambers misses an opportunity. While the untimely is varied, disruptive, and contingent, it might be patterned or outlined, its nebulous form characterized by approximation, or, as Bergson would say, by intuition. The untimely may be spectral, in Chambers's terms, but even Hamlet's father, whose very wraith initiates the out of joint experience, is richly described in Shakespeare. If the untimely experience is of value, either because of its romantic and expansive potentiality or because of its dissonant and painful immediacy, it is worth sketching its form and trying to outline its furtive shape so we can attend to its moments of emergence, albeit without efficient causality.[3] Elizabeth Grosz's *The Nick of Time*, which examines the nexus between time, evolution, and politics in developing a political ontology of becoming, is a welcome corrective in this regard. But Grosz's book, which offers a perceptive and nuanced reading of Darwin, Nietzsche, and Bergson, only offers tantalizing hints of broader political implications that are never fully explored.

Why then is it necessary to explore the political implications of the untimely? What might be problematic about clinging to the dominant image of time in Western thought—a linear and sequential concept of time that moves relentlessly forward toward the future? One way to respond to these questions is to borrow from Judith Butler's sharply critical language about the ossification of liberal principles and apply it to the theorizing of temporality, which is marked by similarly universalizing rhetoric (which is not surprising because liberalism itself is structured by linear and sequential time). Butler writes that such accounts "cannot yield the kind of analytic vocabularies we need for thinking about global interdependency and the interlocking networks of power and position in contemporary life".[4] When it comes to the language of temporality, we suffer from a similarly limited vocabulary. For instance, consider the language of universal progress, which structures and undergirds the linear and sequential concept of time, but which proves to be quite problematic when looked at in a contemporary political context. Not only do the critical assumptions sustaining such concepts then go unexamined, but they continue obscuring productive areas of analysis. One example of such conceptual constraint (which is discussed at length in the Chapter 5) has to do with our analysis of secularism in the Muslim world. When we approach the question of secularism, we pay scant attention to its historical and cultural roots in a Christian past and assume that the secular concept of time structuring much of Western politics and theory can be easily applied to the Muslim world. When such foundational assumptions do not translate in the face of cultural difference and disparate histories, we resort to the language of anachronism, where the Muslim world is quickly disparaged as backward and as refusing stubbornly to move forward into modernity.

The concept of time that we use as an analytic lens remains hopelessly smudged and obscures the "networks of power and position in contemporary life" as we cling to a universalist model that has limited applicability when shorn of its particular context. This makes the following question posed by Wendy Brown an urgent one to consider as we try to unpack the political implications of clinging to one image of time. Brown asks, "When fundamental premises of an order begin to erode, or simply begin to be exposed as fundamental premises, what reactive political formulations emerge—and what anxieties, tensions or binds do they carry?"[5] In order to explore this question, this chapter draws on Sheldon Wolin's succinct essay "What Time Is It?", which brings to the foreground the relation between political theory and a particular concept of time. Wolin argues that in our increasingly globalized and diverse world, there is no longer a single shared "political time" but that there are instead culturally constituted, fragmentary, and distinct times. The effect of such a splintering of a previous unity is to disrupt common narratives of what constitutes the political. In order to explore the consequences of this splintering, let us look closely at an account of an interrogation at a U.S. military base in Afghanistan, during which multiple and incommensurable concepts of time clash in a moment of fatal suspicion. This incident serves to underline the saliency of Wolin's problematization of "our time", a singular and shared conception of time, and, in doing so, gestures at the necessity of reworking notions of temporality within political theory or at least opening it up as a site of contention. In contrast to Wolin, and to some extent to Brown, both of whom decry this temporal fragmentation and use it as an opportunity to plead for a more deliberative and quiet time as more consistent with a certain concept of political thinking, I argue instead that a complex image of time is essential to the theorizing of politics in a globalized world of difference.

THE SPLINTERING OF "OUR TIME"

Sheldon Wolin's concise article "What Time Is It?" is a layered and thoughtful response to a concern posed by Jeffrey Isaac—a concern with theory's inability or unwillingness to react to dramatic events of "our time", such as the overthrow of Soviet style regimes in Central Europe.[6] Wolin's response to Isaac is particularly useful not just because it attempts to call into question theory's relationship to the event but because at its core is the problematization of the concept "our time". Wolin's interest in what he calls the "language of temporality" is driven by the concern that political theory is particularly difficult today, a difficulty he attributes to the "instability of political time".[7] This arises from a certain theoretical sensibility, which while it recognizes the dramatic character of epochal events, such as the fall of the Berlin Wall, objects to the formulation "our time", with its implication of a homogeneous, shared time. The objection is in part that there is no single shared "political time", only culturally constituted different times".[8] Wolin's sharp observations give rise to a set of important questions that are essential for exploration in contemporary politics. What is "our time"? Is it a concept so intuitive or self-evident that it requires no further clarification? What does it mean to have a common time and who shares in it? Is such a notion of time even possible today?

"Our time" is indeed a familiar time: it is the temporal narrative that has dominated Western thought for centuries and that has provided the metaphysical skeleton for its most significant theoretical contributions, from liberalism to communism. "Our time" is a progressive and sequential time, linear in form, where the past is succeeded by a present, which is then enveloped by the future. It draws on Christian traditions from which it adopts its metaphorical linearity, a straight line from Genesis to the Day of Judgment. It is in some ways the legacy of the modern—our contemporary understanding of it having less to do with an awareness of the Christian tradition that gave rise to it, than with familiar notions of historical development, inexorable progress, technological advancement, and forward motion generally. It is a concept of time that orders the messy events of our lives into a straight line—an irreversible current, always moving us toward the promise of an end of history, or a more enlightened age, or some sundry teleological end. The notion of such a "shared, homogeneous time", of humanity moving together along the same temporal path, governs much of historical and political understanding, and for political theory, it makes the task of theorizing the "dramatic events of our time" not only less objectionable and more authoritative but also certainly less complicated by the way it ignores the notion that the shared community of "we" with its corresponding "our" is shot through with difference and conflict.

Wolin questions this foundational assumption of "our time", of a shared time, positing in its stead the notion of multiple, culturally distinct accounts of time. He writes, "The objection is in part that there is no single shared 'political time', only culturally constituted different times. Their self-conscious character produces the equivalent of a different time zone that contributes to a disruption and undermines the possibility of a common narrative structure and along with it, a common identity—formerly a staple element in conceptions of the political".[9]

In this formulation, Wolin introduces some key terms that are helpful in evaluating the relationship between the concept of temporality and the political. He observes that "time" is not simply "our time" or the shared assumption of an unproblematic unity. Instead, it appears that there are multiple culturally distinct times. In other words, the increasingly diverse world has problematized previously held concepts of uniformity and in its place yielded multiple and experiences of time that no longer fit into the common narrative. This, however, is not merely an empirical observation. The key point for Wolin is that this splintering of a former unity is the consequence of a certain "self-conscious character", a developing awareness of difference that introduces into the mix "the equivalent of a different time zone". Starting from the premise that the fundamental philosophical concept of "our time" is no longer tenable, he introduces the notion of multiple times in its stead as well as a conscious awareness that produces a different time zone.

What are these diverse time zones Wolin alludes to that are inaugurated by an increasing self-awareness? Are they simply the different zones of time around the globe marked by their deviance from GMT? Or are they, more interestingly, different zones of speed within a splintered concept of time, with slow, agrarian societies on one end and the high-tech, fast-paced cosmopolis of Tokyo on the other? Do they differ in their form, with cyclical concepts of time on one hand and progressive, sequential time on the other? Perhaps they are constituted by the increasing acceleration of technology in modern industrialized nations, or by an attentiveness to the silencing and redemptive powers of the historical past as in genocide ravaged Rwanda and the Balkans, or by a deep attunement to man's relationship with nature's rhythms in the rice paddies of Bangladesh (albeit interrupted constantly by a ringing cell phone). Furthermore, these culturally different images of time must also contain within them multiple experiences of time; in particular, the disruptions and out of joint-ness of the untimely, which continually emerge to jostle their organizing frame. While lacking in precise explication of these concepts, Wolin's formulation is invaluable in opening up a key front of inquiry into the complex relationship between the experience and understanding of time and the theorizing of the political. By introducing a critical element of difference and the cognition of difference it into an otherwise homogeneous narrative, Wolin correctly identifies the dynamic at work in the problematization of the temporal relationship. The disruption of the common temporal narrative, in Wolin's view, leads to the disruption of a common identity, a dynamic that has significant political consequences such that formerly "staple element(s)" now appear contestable. If time is indeed composed of multiple zones of different speeds and forms, how might an encounter between these different zones be resolved and what might it reveal about the contemporary condition?

(MIS)RECOGNITION, POWER, AND TIME: INTERROGATING THE RUPTURE

In order to focus more closely on the disruption of a common temporal narrative and its political consequences, I want to examine a news account of an interrogation of an Afghani suspect by his CIA interrogators and unpack it in some detail. Before I turn to a recounting of the incident, some further clarification is required. The details of the interrogation are disturbing on multiple grounds, and the story itself ended with the brutal torture and death of the Afghani suspect, Wali. In subsequent months, one of the interrogators, Dave Passaro, a CIA contractor and Army Ranger, was put on trial and convicted of felony assault.[10] His sentence was later reduced on appeal.[11] The case raises all kinds of complex issues about the CIA and its interrogation policies and practices and about the significant human rights abuses that were committed and continue to be committed today under the rubric of the "War on Terror" (though we now avoid using the term itself). I want to make clear at the outset that while those are gravely important issues, I am less interested here in exploring them than in highlighting a moment of temporal incommensurability in the interrogation itself, one that gives rise to intense suspicion. I also want to note that I am less interested in whether the suspect, Wali, is guilty of firing rockets at the army base or of traveling to Pakistan to meet with the Taliban. For my purposes, the veracity of his account is not important, and I am not interested in whether his inability to participate in his interrogators' temporal narrative is based on authentic experience, or simply a refusal to participate. Even more than the temporal incommensurability that is signified by this account, what is of interest here is how the encounter itself highlights the difficulty of recognizing others when operating within a hegemonic frame, the need for a complex temporal vocabulary to call attention to these operations of power, and how the insistence on one common account of "our time" is an exercise of power in itself as this account brutally illustrates. Turning once more to Butler, "part of the very problem of contemporary political life is that not everyone counts as a subject. Multiculturalism tends to presuppose already constituted communities, already established subjects, when what is at stake are communities not quite recognized as such, subjects who are living, but not yet regarded as 'lives'".[12] In this account, the problems of recognition, or of misrecognition, as Patchen Markell would put it, are profound, and they are inaugurated by multiple, clashing, and dissonant images of time.[13]

The following account of the questioning of an Afghani man by CIA contractors was revealed by the CIA interpreter who was present for the interrogation in an interview with the *New York Times Magazine*. The interrogation took place at an American military base in Afghanistan where Wali, a native of Kunar, was questioned by the Americans about a possible trip to Pakistan that might reveal Taliban connections. This brief excerpt of the exchange is revealing in that it brings to light the differing temporal narratives, or the different zones of time, to which Wali and his interrogators seem to be subscribing. Note how the incommensurability of these two temporal structures, or the "splintering" of the temporal narrative, results in both misunderstanding and suspicion and, in particular, how Wali's inability, whether authentic or not, to answer questions that center around a common temporal structure and instead put forward a different image of time, can simply not be recognized under the terms of the hegemonic narrative to which his interrogators subscribe.

> *June 2003. Interrogation at American base near Asadabad, Afghanistan. Afghan with flowing hair named Abdul Wali. 3 days later, Wali, 28, was dead and a CIA contractor was accused of beating Wali to death using his hands, his feet, and a large flashlight.*

> The interrogation started casually enough. Have you been in contact with Taliban? Were you Taliban? The inquiry then turned to Wali's recent visit to Pakistan. "How long ago were you in Pakistan?" Wali looked confused, and I doubted he'd be able to answer. People in Kunar don't have calendars; most of them don't even know how old they are.
>
> BRIAN: "You don't have to give a specific date. Was it two, three days ago? Two, three weeks ago? Two, three months ago?"
> WALI: "I don't know. It's really hard for me to say."
>
> The Americans exchanged glances. I prodded him: "Can you at least say a week or two weeks or a month, or two months or something?". But he couldn't. For him as for many of his countrymen, time unfolded forward—there was no way to go back later and try to fix it in a structure.
>
> WALI: "I just, I go to sleep, I wake up and there's a next day. I feed myself, I go to sleep and there's a next day."
>
> The Americans weren't buying it. Dave took over the questioning. Even seated, he seemed enlarged by anger . . . "If you're lying, your whole family, your kids, they'll all get hurt by this", he threatened.[14]

This exchange begins with a mild query about a possible trip to Pakistan that seeks merely to gather information. Even Wali's initial confusion is treated in a helpful manner—"you don't have to give a specific date", Brian says. Any answer within this frame of reference will suffice, he implies, without a trace of self-consciousness. By the end of the exchange, however, the tone is considerably different; not only does the interrogator change, but the questioning is also inlaid with suspicion and takes on a threatening quality. The next we hear of Wali, he is dead, the result of brutal torture, and an inquiry has been opened into the matter. Why is this interrogation significant? It is of interest because the dramatic shift seems to hinge on a moment of temporal confusion, at which different experiences of temporality collide. What accounts for the dissonance is Wali's inability to respond to an apparently innocuous question—"How long ago were you in Pakistan?" In the eyes of the American interrogators, Wali appears to suffer from a temporal disability, an inability to account for his experience in familiar temporal terms. The temporal narrative of his American interrogators, a time image that is linear, sequential, and forward moving—structured as Brian says by quantifiable, successive units of "two, three days ago? Two, three weeks ago? Two, three months ago?"—does not seem to translate for Wali. He does not appear to know how to respond to an inquiry structured by a concept of time whose rules seem to be foreign to him (at least in the interpreter's recounting). He struggles to explain his temporal perspective in which, in this third person account, he seems to argue that time is not fixed into an easily quantifiable structure—"I just go to sleep, I wake up, and there's a next day". His response seems to his interrogators to be evasive and dismissive, as it is refers to a an image of time different from their own, where time is not something relentlessly moving forward and quantified in precise units but is something that flows in a slow rhythm of repetition. His suspicion aroused, the interrogator is provoked by this moment of seemingly culturally constituted temporal dissonance and quickly becomes angry, even threatening. Wali's inability to express himself in the interrogators' familiar temporal vocabulary introduces into the interrogation a note of the fatal. Wali is bludgeoned to death. While we do not know precisely why, this exchange illustrates the splintering of "our time" in an all-too-human fashion. Splintered time, where the expectation of a "common narrative" splits like so many shards of wood into alternative experiences of time inaugurates a powerful and disciplining event. It silences Wali.

But there is more to this exchange than the silencing of the other. It appears to be an encounter between diverse zones of time that vividly illustrates the rupture diagnosed by Sheldon Wolin. The moment of suspicion, aroused by a failure to recognize another's temporal experience, underlines the apparent incommensurability of "culturally constituted different times" and the disciplining moves required to hold together a single unifying narrative of time. Wali's silencing provides an extreme illustration of the political consequences of the splintering of a common narrative of "our time". It also clearly underlines what Patchen Markell has diagnosed as the problem of recognition, or indeed of misrecognition. For Markell, the "ideal of mutual recognition, while appealing, is impossible, even incoherent . . . in pursuing it we misunderstand certain crucial conditions of social and political life".[15] In aspiring to mutual recognition, we too easily ignore our vulnerability to others and the structures of power that sustain our interactions. Thus, social and political subordination does not emerge from systematic failures to recognize the identities of others but rather from "ways of patterning and arranging the world that allows some people and groups to enjoy a semblance of sovereign agency at others' expense".[16] This is precisely the dynamic that underlines the encounter described earlier, where the privileging of a universalist conception of "our time" by the CIA interrogators and the sovereign agency it affords them is sustained at Wali's expense and utter vulnerability to power. It is not that Wali is not recognized because of the failure of an abstract liberal ideal of mutual recognition; rather, Wali's misrecognition gestures at something much deeper about the way in which power relations structure our encounters by allowing some to act with sovereign autonomy while stripping it away from others. What is notable about Wali's encounter with the interrogators is how the radical differential in power relations is also shaped by the privileging of an image of time. The dominant concept of time does problematic work here by asserting its sovereignty and refusing to recognize the legitimacy of other perspectives. In his work, Markell notes the role that temporality plays in the failures of recognition but does not pursue it nearly far enough. He writes, "I cast the problem with the ideal of recognition as in the first instance a problem with its temporality—that is, with the way in which it expresses the aspiration to a sort of sovereign invulnerability to the open-endedness and contingency of the future we share with others".[17] Markell correctly identifies that such aspirations do not adequately take into question the contingency of the future, but he fails to note the deeper sense in which they also fail to note the contingency of the dominant concept of time itself. Thus, not only is the problem of political recognition an attempt to assert sovereign agency at the expense of others, but, as illustrated by Wali's interrogation, it is also sustained by a universalist temporal narrative that cannot recognize alternative accounts and is closed off to contingency of time as experienced out of joint.

POLITICAL TIME, SPEED, AND EROSIONS

Given his sensitive diagnosis of the rupture of a common temporal narrative, one might expect Sheldon Wolin to provide further theoretical insight into the political implications of such a break in "our time". Interestingly, for Wolin himself, such a diversity of temporal narratives helps to promote "the instability of political time", and such a plurality of temporalities makes the task of theorists "daunting".[18] Surprisingly, it would appear that for Wolin, the brutal interrogation described earlier might serve only to illustrate the chasm between multiple culturally constituted times by creating a condition of friction and confusion that makes democratic politics impossible. Indeed, the rupture in "our time" that Wolin so sharply diagnoses does not seem to present for him a new terrain of opportunity where foundational concepts can be reexamined and alerted to both new dangers and possibilities. Instead, it is a "daunting" prospect that makes the task of theorists particularly difficult. For Wolin, the activity of politics requires not only shared understanding, but deliberation and slow speed. He writes, "Political time, especially in societies with pretensions to democracy, requires an element of leisure . . . in the sense, say, of a leisurely pace".[19] In this formulation, not only does he restrict his inquiry to "political time", as distinct from cultural, evolutive, or natural time, but political time is also conflated with the notion of a slow and leisurely pace. Wolin's response to the splintering temporal narrative appears to be to marginalize and isolate political time from its disruptions, a separation that is questionable at best and that betrays a desire to keep political time isolated from the fluxes and flows that seem to characterize much of late modern life or, as Markell noted, a sovereign invulnerability to the contingency of time. Such a desire to preserve or, more sharply, fossilize political time also appears to rely on a particular notion of politics, where the political realm is configured as one that moves slowly in terms of pace and uniformly in terms of progression to make consideration and tranquility possible. As such, Wolin's response is a nostalgic one, limited by a narrow and protective understanding of political time that privileges the "leisurely pace" of deliberation as central to political activity, which leads him to the conclusion that political theory, amid a rupturing temporal narrative, is particularly "difficult today".[20]

Beneath Wolin's characterization of democratic politics set in slow time lurks a model of thought that seeks to quell the disruption and the emergence of the new that characterizes the out of joint experience of time. He writes that political theory might "be said to be governed by political time . . . in the deliberations about political life that figure in each and every theory and make their construction such a slow and drawn out process".[21] Wolin does not provide a definition of political time, how to distinguish it from other zones of time, or what its limits might be or what disruptions it might contain. The one characteristic that seems to identify political time is its slow and deliberative pace. This understanding of political time, and consequently of political theory, as being limited to zones of slow speed and tranquil deliberation seems to altogether overlook the accelerated pace of late modern politics, with its twenty-four-hour cable news and social media networks—an age of rapid response that invites spontaneous deliberation from a plurality of perspectives and thrives on friction and flux.

Given this perspective, it is unsurprising that Wolin finds the undermining of the common temporal narrative problematic and daunting, as he seems to covet a mode of thinking that requires a deliberative past and that implicitly stills new and out of joint perspectives. However, politics, in its vivid, real, conflict-ridden sense may not always be consonant with his invocation of tranquil deliberation. While Wali's interrogation exposes dangers in the experience of multiple zones of time set at different speeds, it may be that democratic politics itself requires such a condition to flourish. As Wolin concedes later, "political time is conditioned by the presence of differences and the attempt to negotiate them".[22] For him, however, the presence and the negotiation of these often profound differences "preserve time".[23] Is the consequence of the negotiation of political difference necessarily a fossilizing one? In the preceding case, the result of such negotiations is not a frozen and tranquil state but is instead a brutal death possibly triggered by the clashing of multiple accounts of time, where one account cannot be recognized by the other. Differences in temporal narratives and their negotiation therefore cause friction that is real and tangible. Can the presence of this fractious space of negotiation, however, be a space not merely for the exercise of a repressive and disciplining power, but perhaps also the site of a productive power that forges new assemblages from its disturbances? Put another way, perhaps this undermining of the common thread, this splintering of "our time", can create the possibility for the construction of a complex temporal narrative that is open to disruption and multiple zones of speed, which allows for a fuller and more critical understanding of contemporary politics.

If the contemporary world is a fragmented space full of shifting and contingent relationships that undermine established conceptions of sovereignty and identity, then a more complex understanding of time as splintered and multiple may be more suitable for theorizing the political as it exists today. It is revealing that even as Sheldon Wolin argues that political time requires a commonality and stability that is rupturing globally, he locates diverse temporal zones elsewhere in the social sphere in the realm of culture and economics, which for him remain clearly distinct from, and in contrast with, the political. Wolin writes, "In contrast to political time, the temporalities of economy and popular culture are dictated by innovation, change and replacement thorough obsolescence. Accordingly time is not governed by the needs of deliberation but by those of rapid turnover".[24] While Wolin correctly identifies the accelerated pace of living, he draws a taut and unsustainable division between culture and politics, where cultural and economic time are fast-paced zones of flux, and political time, which is the slow time of preservation and deliberation. How can such a sharp contrast between politics and culture and economy possibly be sustained? As Wolin acknowledges a little later in his essay, "It is clear today that culture and economy have both so thoroughly transformed politics that it becomes difficult to recall when they did not".[25] Wolin is himself aware that politics cannot be isolated and separated from the realm of culture, but he draws a clear and unsustainable binary between slow, static political time and fast, changing, cultural time.

While Wolin's account of zones of cultural and economic time is shrewd, his insistence on separating culture from politics prevents him from the additional step of complicating political time and therefore closes him off to the possibility of a more layered understanding of contemporary politics that embraces a concept of time that is out of joint, disruptive, and composed of multiple zones of speed. Indeed, the very conflation of the cultural and economic with the political that Wolin decries can occasion the grounds for a rethinking of the political and the role that temporality plays in many of its formulations. Wolin, however, remains useful for the purposes of this inquiry because he highlights the political dimensions of temporal experience by diagnosing the splintering of the common narrative and the multiple culturally constituted zones that appear to be emerging in its place. He may even be correct in asserting that political time is slow, deliberative, and distinct from other temporal zones, but he is limited is in his refusal to consider the possibility that such difference may yield productive outcomes for democratic politics.

More helpful in this regard is Wendy Brown, who, in *Politics Out of History*, begins to probe the space that emerges from such a loosening of the time and politics relationship. Brown is particularly concerned with the political and ethical consequences of our unmooring from foundational attachments and insistently asks, "What happens when the beliefs that bind a political order become fetishes?"[26] The attachment to a common temporal narrative that once girded political understanding is transformed from the status of foundational premise or belief to mere nostalgia or, more problematically, is fetishized longing in the face of its splintering unity. A fervent desire for this former unity, in Brown's view, can engender a deep sense of pessimism, or as Wolin put it in the essay outlined earlier, a sense that "theory is in a bind".[27] Brown suggests that the emerging self-consciousness, which Wolin highlighted as a key component in the splintering of narrative unity, by its very recognition of this rupture dissolves into despair. She writes, "When a disintegrating political or cultural narrative seems irreplaceable, panicked and reactionary clutching is inevitable; when this perceived irreplaceability refers to a narrative or formulation actually lost, melancholy sets in".[28] Brown's diagnosis of this condition seems to echo Wolin's, in terms of the extent to which they both perceive the loss of such a foundational premise. While Wolin's reaction is one of nostalgia, Brown's is concerned with the hopelessness that such a loss engenders and the "despair about the capacity to grasp our condition and craft our future".[29]

Brown's language is pessimistic as she charts the terrain of this crisis of belief, but unlike Wolin she sees some productive potential in this fracturing. She cautiously suggests that "the same features suggest in the present a porousness and uncharted potential that can lead to futures outside the lines of modernist presumptions".[30] As she acknowledges herself, she is "neither purely despairing nor purely hopeful", and her response to the condition she diagnoses "bears the mixture of heaviness and hope".[31] The dual observations that Brown makes about the experience and reactions to such a condition of erosion are essential as they provide a set of lenses through which to examine the political and ethical consequences of destabilizing the common temporal narrative. For now it suffices to note her cautious diagnosis of the condition and to attend to the potential in the rupture that she identifies for theorizing the political. The succeeding chapters will make the case that political theory need not be in a bind amidst the splintering narrative of "our time". Rather, political theory can incorporate within it a complex account of time that will make it open to the constructive frictions, the agonistic relationships, and the productive tensions that permeate and complicate our concepts of the political.

In this fragmented yet globalized world of a dizzying variety of identities, theory may not only be possible; indeed, theory may thrive amid such out-of-joint, disruptive, and multiple accounts of time which simultaneously complicate our understanding of politics and open up the possibility of examining the contemporary condition from different angles. It is therefore useful at this point to turn away from the diagnosticians of the rupture to the splintering itself, to lay out the defining terms of the common temporal narrative which is fracturing, and to explore some alternative concepts that may take its place. As Brown writes, while we may have "lost confidence in a historiography bound to a notion of progress or to any other purpose, we have coined no political substitute for progressive understandings of where we have come from and where we are going".[32] It is to an exploration of this temporal landscape, which is both dominant and where we have come from and which may provide an alternative perspective and suggest where we might be going to which this book now turns.

NOTES

1. See Samuel Chambers, *Untimely Politics: Taking on the Political* (New York, New York University Press, 2003). William Connolly, *Neuropolitics: Thinking, Culture, Speed* (Minneapolis: University of Minnesota Press, 2002) and *A World of Becoming*. (Durham, NC: Duke University Press, 2011). Elizabeth Grosz, *The Nick of Time: Politics, Evolution and the Untimely* (Durham, NC, and London: Duke University Press, 2004).
2. Chambers, *Untimely Politics*.
3. In his recent work, Chambers has distanced himself somewhat from his earlier book by pushing the analysis further toward the social formations of the untimely, which is both necessary and significant. However, even as he dismisses what he now calls a binary between nonlinear time and clock time, he continues to pay short shrift to the experience of time, often reducing it to a crude empiricism without subjecting it to any kind of layered analysis. See "Untimely Politics Avant la Lettre: The Temporality of Social Formations", *Time & Society* 20 (July 2011): 197–223.
4. Judith Butler, *Frames of War: Precarious Life, Grievable Life* (New York: Verso, 2009), p. 31.
5. Wendy Brown, *Politics Out of History* (Princeton, NJ: Princeton University Press, 2001), p. 3.
6. See Jeffrey Isaac, "The Strange Silence of Political Theory", *Political Theory* 23 (1995), no. 4: 636–652.
7. Sheldon Wolin, "What Time Is It?", *Theory and Event* 1, no. 1 (1997), http://muse.jhu.edu/journals/tae.
8. Ibid.
9. Ibid.
10. Scott Shane, "C.I.A. Contractor Guilty in Beating of Afghan Who Later Died", *New York Times*, August 18, 2006, http://select.nytimes.com/gst/abstract.html?res=F20910FE3E5A0C7B8DDDA10894DE404482&emc=eta1.
11. Matthew Burns, "Sentence Reduced for CIA Contractor Convicted of Detainee Abuse", April 6, 2010. www.wral.com/news/news_briefs/story/7366849/
12. Butler, *Frames of War*, pp. 31–32.
13. Patchen Markell, *Bound by Recognition* (Princeton, NJ: Princeton University Press, 2003).
14. Hyder Akbar, "*Memoir: Interrogation Unbound* as told to Susan Burton", *The New York Times Magazine*, Section 6, July 11, 2004.
15. Markell, *Bound by Recognition*, p. 4.
16. Ibid., p. 5.
17. Ibid., p. 15.
18. Wolin, "What Time Is It?"
19. Ibid.
20. Ibid.
21. Ibid.
22. Ibid.
23. Ibid.
24. Ibid.
25. Ibid.
26. Brown, *Politics Out of History*, p. 4.
27. Wolin, "What Time Is It?"
28. Brown, *Politics Out of History*, p. 4.
29. Ibid., p. 5.
30. Ibid., p. 5.
31. Ibid., p. 17.
32. Ibid., p. 3.

2 "Our Time" and the Transcendental Image of Thought

What is "our time", the time that is so familiar it requires little or no explanation? What are the contours of this shared and homogenous time-image, which, in the language of Sheldon Wolin provides the common narrative structure for theorizing the political? To begin to explore this temporal landscape, it is essential to examine what is being considered as a fundamental premise, namely, the dominant and teleological concept of time. In the history of political thought, indeed in the central works and ideas of Western civilization, it has retained hegemony despite challenges to its authority. It assumes that events follow each other sequentially, that they move purposefully towards a defined end, and that they are powerfully motivated by the modern concept of progress or historical development. In a concise essay titled "Time and History", Georgio Agamben provides a succinct account of this image of time. The competing concepts that emerge from a teleological image of time are cyclical and linear time, both of which are sequential in nature. The Greeks, from Aristotle onward, put forward a circular or cyclical notion of time, central to which was the idea that all human affairs have a natural movement, a movement that is circular and repetitive, which results in a continuous standard rotation. Cyclical time, as the name suggests, is an image of time that is circular in form, and purposive and teleological, but it is also a sequential concept where the present gave way to the future which then gave way to the past. It is a stable, closed, system where each temporal stage folded seamlessly into the next, with no room for disruption or splintering, guided only by its telos, the inevitable end point toward which it unfolded. Cyclical time, with its sequential character and purposive nature, provided the foundation for what became the dominant account of time in Western civilization: linear and progressive time.

The linear notion of time deviates from the cyclical image of time in important ways even as it inherits from the latter its sequential nature. For instance, one of its dominant features is that it also sees time as an infinite continuum composed of quantified instants—past, present, future; morning, noon, evening. Its spatial model, however, is markedly different. Being linear in form, it replaces the circle with a straight line, such that instead of operating in a repetitive loop, time now insistently proceeds forward, in a single direction, without deviation and without return. It is an irreversible image of time but it is its linear shape that provides its distinctive character. It is a "structured process in terms of before and after" where past flows into present which flows into the future, where the past never returns, where each fleeting instant gives way to the demands of an impending future.[1] Linear time hurtles forward; it is pure, sequential chronology, which can only sustained by some concept of continual, infinite progress toward a divine or a human end.

This linear account of time is Christian in origin and therefore travels purposefully in one direction from Genesis toward the Day of Judgment, its inevitable end. Even as the linear notion of time is separated from Christianity and gradually secularized, it retains its teleological focus on the future, replacing the Christian faith in salvation that previously drove it with a strong belief in a tendency toward progress, toward realizing scientific and moral potential. This faith in progress is driven by the ascendancy of reason and rational thought in the Enlightenment and the Industrial Revolution, in which the future came to be viewed as an unmapped terrain that could be illuminated by the beacon of reason. The future thus came to be seen as a vast source of potential that could be realized by technological and rational advancement, rather than as some preordained end at which one merely arrived. The transformation from a Christian and linear time-image to a secular and modern one recharacterized the progressive image, shifting it from divine providence to a postulate of moral life itself, one to be deciphered, refined, and realized by reason. The assumption of linearity that is retained in this secular and progressive time-image gives the instant an ephemeral and fleeting character—the past does not coexist with the present, and the present gives way to the relentless demands of a future that constantly impinges on it. This is the sequential and progressive time that our clocks and watches structure and enforce every day, during which each second ticks and is replaced by the next.

This chapter traces the emergence and influence of this linear and progressive concept of time that has dominated the Western philosophical tradition by looking closely at the writings of two influential thinkers: Augustine and Immanuel Kant. Both Augustine and Kant reflect on the temporal conditions of experience and are of particular interest because they both write during periods of transformative change—when new faiths are emerging, when fundamental premises of the old order are eroding, and when foundational concepts are called into question. Each stands at a crucial moment in history—Augustine at a time when the Christian religion is spreading and the vast Roman Empire is in decline and Kant on the cusp of modernity, during the Age of Enlightenment, when reason is challenging faith. Both are engaged in attempts to theorize the transformative changes of their present, and in doing so, they succeed in articulating a powerful time-image that has dominated the Western philosophical tradition.

For Augustine, lived time is a dizzying confusion that must be stilled and transcended, and this image of time shapes his edict to purge the soul from passions, from lust, to confess sins into being that can be healed by a creative God in his eternal tranquility. Such an image of time, which requires confession and conversion and faith to quiet disruption, is a transcendent image of eternal stillness that seeks to impose order and quiet on the disorder of living and to replace it with the placidity and stillness provided by faith in a transcendent God. For Augustine, this image of time is inextricably linked to the problem of evil and its diagnosis from the Fall to the divided will that can only be healed by grace, by the progressive realization of redemption. In its transcendent nature, in the silencing and disciplining of the body that manifests in the division of the will, and in terms of its eschatology that sees time as called into being by Creation and ended by Armageddon, it is Augustine's rendering of his Christian faith that is determinative for his construction of time.

For Kant, an engagement with time is also one of the most pressing questions of his age. Kant epitomizes not just the strains of the Enlightenment but also the anxiety of the modern and seeks to reconcile the notion of a distinctively new age that often refers to itself as an epochal break from the past with a strong desire to retain faith and morality and re-refashion it for his time. Kant reaches toward the linear notion of time that Augustine articulates but seeks to revise it for modernity while still wrestling with many of the same concerns of his predecessor—the problem of evil, the moral imperative, and the freedom of the will. Kant's account of time requires faith, not a faith explicitly in an omnipotent God but a faith that is reworked for the modern age, a faith that is enforced by the moral imperative and by the regulative laws of reason and the idea of progress and that wears its transcendental structure more lightly. Kant's writings on the Enlightenment bear the imprint of the anxiety of the modern, the fear of losing the moorings of faith, the dread of the misuse of reason, and a general apprehensiveness about the vagaries of sentiment.

This chapter puts Augustine and Kant in conversation with each other to explore the dominant image of linear time and the image of thought it engenders. Its purpose is twofold. First, it traces the development of linear and progressive time from its Christian to its secular and modern form and looks at the extent to which the latter is both the legacy of the former and a revision of it. To that end, it explores how Christian faith is revised in the modern age by the assumption of a law-like model of nature and how a future whose end is ordained by God is changed into one that is open-ended and replete with possibility while retaining in it amid not just the fundamental contours of linear time but also a categorical moral imperative that preserves religion with the limits of reason. Second, it contends that such a concept of time tends toward an intellectualist image of thought, defined as a mode of thought that turns away from the body, which it perceives as corrupting and disquieting. Such an image of thought attempts to move beyond the disruptions of sensory experience to the tranquil life of the mind, by appealing either to a transcendent authority or to regulative laws of understanding. It argues that Augustine and Kant, both of whom recognize the significance of temporality and wrestle with its questions deeply, come away from their engagements with the impulse to quell the anxiety that the temporal question provokes by appealing instead to a transcendent being or image of thought.

FROM CYCLICAL TO LINEAR TIME: THE CHRISTIAN TRANSFORMATION

The notion of time as progressive—moving sequentially from past to present to future—and as linear—taking the shape of a straight line—is at the heart of Western philosophical tradition. However, in the history of Western philosophy, this now-familiar linear concept of time once represented a radical break from the ancient tradition. As portrayed in Aristotle's *Physics*, the classical concept of time was a cyclical and stable system. Time was conceived as being made up of cycles of repetition during which the lessons of the past would be repeated and the future was not perceived as a source of disruption that could transform the fundamental order of things. Aristotle acknowledged a discernible development, or *physis*, in the history of human experience but did not entertain the possibility of future progress and transformative change. In the ancient tradition, the cyclical time of repetition emphasized above all the constancy of human experience.[2] This static and cyclical concept underwent a dramatic change with the advent of Christianity and the rising influence of the church during the declining years of the Roman Empire. The Christian Church recast the notion of temporality from a stable cyclical system of repetition to a linear model, with a clearly defined beginning and a fixed end. According to biblical teaching, the world and therefore its time began with Creation and would end on the Day of Judgment as revealed by God's word and affirmed by his prophet Jesus Christ. This new theological doctrine engendered "a shift in the idea of time and history . . . whereby the ancient understanding of the past gave way to an understanding of the past predicated on a belief in creation".[3] As John Pocock tells us, the Christian model, in contrast to the ancient cyclical model, "repudiated all ideas of cosmic recurrence, 'the wicked dance in circles'; such a vision of things would make the world uncreated and endless".[4] The Christian belief in an omnipotent God who had created the world and would bring it to a defined end made the ancient notions of an endlessly repeating past heretical.

What then were the key features of this linear concept of time as it emerged in the early years of the Christian Church? The first and most important feature is the "Christian insistence on a God who had created the world and men at a point in time past and would redeem men and end the world at a point in time future".[5] Christian time was therefore clearly demarcated as the span between Creation and Armageddon. In addition, the time between these two points was not static or repetitive. It was teleological in nature and had a clearly defined end toward which it progressed. As Georgio Agamben writes, "In contrast with the directionless time of the classical world, this time has a direction and a purpose: it develops irreversibly from the Creation to the end, and has a central point of reference in the incarnation of Christ, which shapes its development as a progression from the initial fall to the final redemption".[6] This progression from fall to redemption marked a radical departure from the classical concept of time as it inaugurated a future that could be utterly transformative, indeed even redemptive.

Time was no longer the universal cycle of repetition, seen in the motion of heavenly bodies; it offered final redemption from the sins of an earthly life and the possibility of tranquility beyond it. As Pocock points out, "The advent of the savior monotheisms . . . reorganized and transformed time by making it an aspect of events whose significance was in eternity. God had covenanted with men, and the covenant would someday be fulfilled . . . God had begun action intended to bring about his redemption, and this process would be at a point in time to come be carried out to its full completion".[7] It yielded an image of time that was fixed in its spatial model, a straight line progressing from Creation to a fixed and determined end that offered the hope of future redemption. The development of the human race was seen in "the light of God's plan as laid out in scripture" and imbued with great religious significance.[8] The history of humanity thus appeared as a "history of salvation, the progressive realization of redemption whose foundation is in God".[9]

This concept of time was therefore radically transformed from its classical understanding by the early Christian Church. It took on a new spatial model (it was now defined as a straight line from Creation to the Day of Judgment), and it no longer focused on the past as a source of wisdom whose lessons would be eternally repeated. Instead, it affirmed the possibility of change in the future and strove for redemption in the eyes of God. Christian time was therefore linear, contained a progressive model of change, and was teleological in its pursuit of a redemptive end. Such a major recasting of the image of time from the ancient cyclical type to the Christian linear model also opened significant space for philosophical reflection. By infusing human experience with the cleansing possibility of redemption, it engendered new modes of thinking about the meaning and the significance of everyday life, the importance of faith and confession, and the temptations of sin. In short, the Christian image of time had a transformative effect on European philosophy—it triggered a break from the classical tradition and its foundational natural concepts and, instead, placed great emphasis on the individual and the significance of his actions as viewed from a yet to be determined future end.

TIME, ETERNITY, AND THOUGHT: THE CONFESSIONS OF AUGUSTINE

The most influential of these early Christian thinkers was Augustine, whose reflections on time shaped much of medieval and early Renaissance thinking on the subject. Augustine's writings on time are fascinating to us today because they deal with the confusion of temporal experience in a deeply personal. He is disarmingly honest about the mysteries of time that haunt him incessantly and that remain a source of confusion and puzzlement to him. This inner turmoil eventually compels him to leave these mysteries behind for the transcendent stillness of faith in God that is forged through acts of confession. In Augustine's writing, amid the searching and the vertiginous despair, there slowly emerges a particular image of thought that can impose order on the unruly desires that mark the human condition. What comes out of this intensely human series of reflections is a transcendental image that urges a move away from the carnal temptations of the body to a less worldly and imperishable place, ordered and stilled by faith in God. His temporal understanding is strongly influenced by his Neoplatonist education, his struggle with faith, and his ultimate conversion, all of which converge in his desire for transcendent truth.

The one Augustine first encounters in the *Confessions* is one who is dissatisfied with his life. The pursuit of pleasure no longer satisfies him, but he is not ready to leave the distractions of the carnal life behind him. He is still bound to his earthly past and is fumbling in search of higher meaning and the eternal time of the creator. He is not yet ready for the stillness and tranquility of an imperishable God, because he considers himself to not yet be fully infused with the Christian faith. In a vivid moment he writes, "In love with loving, I was casting about for something to love; the security of a way of life free from pitfalls seemed abhorrent to me, because I was inwardly starved of the food which is yourself, O my God".[10] The young Augustine is restive but not yet ready to abandon the pleasures of the sensual life. However, even as a young scholar he betrays a desire for a higher source of truth and a strong sense of the insufficiency of ordinary life for intellectual fulfillment. He writes, "O Truth, Truth, how the deepest and innermost marrow of my mind ached for you . . . O Truth, in whom there is no variation, no play of changing shadow . . . all they set before me, were dishes of glittering myths . . . I derived no nourishment from them, but was left the more drained".[11] The Neoplatonist predisposition that Augustine shows in his youth, his search for a Truth without variation, for stillness and tranquility over the play of changing shadow, influences both his later understanding of his faith and his philosophical reflections on Time and Eternity. Augustine has a strong desire to go beyond the commonly held accounts of time as the movement of heavenly bodies. He tries to pierce the impenetrable secrets of time and, in doing so, displays a desire to transcend its variation and its changing shadow by insisting on separating worldly flux from the stillness of Eternity. Augustine's reflection on Time and Eternity therefore begins the process of inlaying the linear Christian model of time with a profound transcendental impulse.

Augustine begins his reflections on Time with the moment of Creation, the Word of God that created the world. He writes, "Your voice made itself heard through the movement of a created thing which was the temporal expression of your eternal will".[12] For Augustine, the human world is but an instantiation of God's eternal will, a temporal expression of his divine plan; God created the world, and he will end it. Our time and history are fixed by his creation and ordered according to his will. However, there is a significant difference between the human time created by God, the Christian time that begins with Creation and ends with Armageddon, and God's time. God is omnipotent and immutable; thus, it is impossible to conflate his eternal time with changing human time. It is therefore imperative to distinguish between Eternity, which is the temporal realm of God, and Time, which belongs to human experience. There is succession in human time but not in God's eternity, which must be immutable. He writes, "In eternity nothing passes, for the whole is present, whereas time cannot be present all at once . . . Who shall take hold of the human heart, to make it stand still and see how eternity which stands firm, has neither future nor past, but ordains future and past times?"[13] Eternity is unchanging, but it is the eternal word that ordains and creates future, present, and past times in the human world. In the human world, time is circumscribed by God, it is decreed by his word. Augustine writes, "[E]verything which begins to exist and then ceases to exist does so at the due time for its beginning and cessation decreed in that eternal Reason . . . This eternal Reason is your Word, who is 'the Beginning'".[14] Augustine captures the contrast between Eternity and historical Time neatly when he writes, "Your years do not come and go . . . In our case, our years will be complete only when there are none left".[15] There is stillness and tranquility in Eternity, whereas in the vertiginous time of lived experience, there is endless change until God brings the human world to an end.

Because Augustine conceptualizes an unchanging Eternity, he is acutely aware that for man the experience of time is one of constant change—he is conscious of "intervals of time", he can measure "passing time".[16] Augustine thus interiorizes the experience of time; that is, he treats the tenses we call past, present, and future as probable realities of the mind alone. He writes, "I have therefore come to the conclusion that time is nothing other than tension: but tension of what, I do not know, and I would be very surprised if it is not the tension of consciousness itself".[17] In locating the experience of time deeply in the self, he conflates it with the temptation and the distraction of the human soul and incomparable to God's Eternity. Having gone deep into the recesses of his mind in his attempt to understand the human experience of time, Augustine reaches a state of confusion. He is confident that such an image of time, which is characterized by flux and change, is separate from the eternal time of the unchanging God. But having separated human time from God's Eternity and interiorized the experience of time as a tension of the mind, he finds he can go no further. He writes, "I have leapt down into the flux of time where all is confusion to me. In the most intimate depths of my soul my thoughts are torn to fragments by tempestuous changes until that time when I flow into you, purged and rendered molten by the fire of your love".[18] The very act of philosophical reflection on time tears at him and is a source of confusion. He is battered by the "tempestuous changes" of time and wants to be calmed by God's grace. Augustine's "anguished interrogation" of time—grappling with the "intimate depths" of his soul tears his "thoughts into fragments", and he cannot comprehend it.[19] He therefore seeks to be absolved of this misery, in the same manner in which he seeks to be relieved of the pain of carnal desire, and to find tranquility in God. His exploration of the tempestuous experience of time only reaffirms his commitment to God—it sets out for him the limits of human knowledge and instills in him faith in the redemptive power of grace—"I will stand still, then, and find firm footing in you, in your Truth which is shaping me to himself and no longer will I tolerate the questions of those people who, sickly under sin's punishment, crave more than they can take in".[20] He seeks transcendence and wants to stop dwelling on the confusions and error of human experience and to put his faith in an eternal God and in his regulative demands. For Augustine, it is sinful to crave more than one can understand, to go beyond the limits of human understanding.

The image of time is one that confounds him with its very slipperiness and fragmentation; it tears at his soul because of its dizzying confusions. Augustine's response to this experience is to still the vertigo, to quell the eddies, and to reach for the eternal time of his omnipotent God, which stands outside the corrupt life of man, stained with original sin. As he struggles with this question, one thought continually appears to him, that in the creator's time, in Eternity, there is no confusion—no slippery present, no yawning past, no expectant future. There is a stillness and a tranquility in Eternity, the time of his omnipotent God. This, then, is the time to which Augustine aspires. He wants to let go of the grasping and fumbling his thought imposes on him and transcend it. The image of time therefore emerges as a key component of the ontological fabric of the Augustinian philosophy because it necessitates a quieting of empirical distractions and questioning so he can achieve the joy of salvation. If human time is a dizzying confusion of constant change that one must look beyond, or rather above, one has to quiet the anxiety and transcend it with the calmness of faith.

Augustine's philosophical reflections on time engender a transcendent impulse that has redemption at its core. If man's experience of time is one of tempestuous flux and distraction, all one can aspire to is to be purged at the end of this time by God's life. The flow of time therefore becomes a deeply interiorized process as the progressive realization of redemption, and as such, it gives rise to a transcendent image of thought that seeks to cleanse the soul of the sins of the body. The quest for eternal redemption requires the purging of lust and temptation and all that is earthly and instead progressing toward a transcendent truth, to that which is unchanging and eternal. Augustine makes this clear when he creates a clear distinction between the impure body and the mind that seeks truth. He writes that "the perishable body weighs down the soul, and its earthly habitation opposes a mind teeming with thoughts".[21] In search of heavenly redemption, he wants to no longer be "hampered by the image of a human body".[22] In this struggle to transcend the body, faith is his guide and confession his weapon. He writes, "I strove with this single weapon to beat away from the gaze of my mind the cloud of filth that hovered around me".[23] This impulse to transcendence is highlighted when he creates a philosophical distinction between the changing and the unchanging, which echoes his reflections on Time and Eternity. He does so without any indecision as he writes, "I unhesitatingly reckoned the inviolable higher than anything subject to violation, and what is constant and unchanging better than what can be changed".[24]

This impulse to transcendence in Augustine's rendering of time, his rejection of the body and its passions in favor of faith and the inviolable and unchanging is deeply entwined with his discussion of the problem of evil and the divided will that are central to both the *Confessions* and the *City of God*. He attempts to discern the human will and the evil attached to it so that a layered self emerges, which reveals sinfulness in its depths. Augustine confesses the free will into being, but the more the will is experienced, the more it reveals its division against itself in its constant distractions by its passions and desires that prevent it from achieving the eternal tranquility of God. For Augustine, there is a moral responsibility to become more alert to the divided will and hence to the problem of evil which is responsible for its division—to discover where it came from and what can be done about it. For this it is necessary for Augustine to go back to the story of Genesis. As Hans Blumenberg writes in *The Legitimacy of the Modern Age*, "The answer that Augustine gave to this question was to have the most important consequences of all the decisions he made for the Middle Ages. With a gesture just as stirring as it was fateful, he took for man and upon man the responsibility for the burden oppressing the world".[25]

In Augustine's interpretation, an omnipotent God created Adam with a free will, and this free will forms the link between God and humanity and the gap in which evil is formed. Here, Adam is the key agent and the significance lies in his free decision to disobey God in his very first act. Therefore, the only time the will was truly free, it was willed to sin, resulting in its later splintering and its division that became its greatest punishment. Adam therefore brought evil into the world at its very inception and, along with it, the stain of original sin, which smears us all. After this fall, only the grace of God rendered through confession and conversion can heal the divide.

Augustine gestures to this division of the will and its connection to this original sin when he writes in the *Confessions*,

> This disintegration was occurring without my consent, but what it indicated was not the presence in me of a mind belonging to some alien nature but the punishment undergone by my own. In this sense, and this sense only, it was not I who brought it about, but the sin that dwelt within me as penalty for that other sin committed with greater freedom; for I was a son of Adam.[26]

Augustine therefore locates the problem of evil within the divided will—a will that was divided against itself when it freely willed the original sin—but he holds out the promise that the will can be healed, by confession, conversion, and grace and by the progressive realization of redemption. Indeed, in the *City of God* he suggests that there is a teleological element to this when he writes, "God was well aware that man would sin and so, becoming liable to death, would then produce a progeny destined to die".[27] However, there is the promise that by his grace such a sin could be overcome and could produce a class of men who could achieve the tranquility of eternity. He writes, "But God also foresaw that by his grace a community of godly men was to be called to adoption as his sons, and these men, with their sins forgiven, were to be justified by the Holy Spirit and then to enter into fellowship with the holy angels in eternal peace".[28] Augustine therefore simultaneously introduces the concept of original sin and inscribes the responsibility for it into the very fabric of man's will and then argues that this division can be only healed by grace and by man entering into fellowship with the angels in eternity. In doing so, Augustine imbues time with both an impulse to transcendence that requires that the passions and the desires of the body be stilled for the tranquility of eternity and an element of teleology in which time becomes the progressive realization of redemption through grace, such that there is an ultimate end to man's sinful beginning in an ultimate embrace with eternity at the end of time.

In an age in which time had been transformed by the Christian Church into a linear model in which the Creation and the Day of Judgment serve as bookends, where it had changed from endless repetition to a stable, progressive line and where it had become the progressive realization of redemption, Augustine's philosophical reflections on temporality are significant. Reflecting on a constantly changing and forward-moving temporality that contrasts with the placid Eternity of an immutable God, engendered in Augustine is an impulse to transcendence that seeks redemption and tries to purge all that distracts from this heavenly calling. It seeks to no longer be hampered by a perishable body in its search of a higher imperishable Truth. Augustine's reflections on a linear and progressive model of time, which throw him into deep turmoil and cause him to yearn for redemption through the unchanging time of Eternity, therefore give rise to a transcendental and teleological model of thought, which seeks to purge the body in the pursuit of higher truth—a model that takes on increasing rationalistic undertones subsequently as Christian time becomes secularized in the age of modernity.

THE AGE OF ENLIGHTENMENT: THE TRANSFORMATION OF CHRISTIAN TIME

Modernity has often been characterized as an epochal consciousness that emerged during the Enlightenment—the product of an age that had severed its ties from tradition and religion and had wrapped itself in the secure blanket of reason. As seen by the philosophers of the Enlightenment, modernity consisted of the "relentless development of the objectivating sciences, of the universalistic foundations of morality and law, and of autonomous art, all in accord with their own immanent logic".[29] Modernity was the distinctively "new" age—it represented a radical break from the past, an unmooring from the foundational concepts of the classical and medieval periods, seeking instead to rework and renew these concepts for its time. Modernity has been defined in various ways—the age in which reason supplanted faith, the age of progress, and the age that above all was gripped by an awareness of the dissolution of traditional forms of life. It has often been regarded as a critical departure from the past because of three monumental events that dramatically changed ideology, religion, tradition, and even temporality—the Renaissance, the Reformation, and the discovery of the New World. The innovative intellectual inquiries of the Renaissance pushed the limits of existent knowledge and called into question categories of established wisdom. The Reformation of Martin Luther brought to each individual the ability to read and interpret the sacred biblical text and threatened the centralized authority of the Catholic Church. Columbus's discovery of an entire New World populated by countless natives literally opened new worlds for exploration. The prevailing sentiment of the modern age was therefore one that embraced the possibility of change, one that was infused throughout with a profound sense of discovery, and one that held tightly to belief in the transformative power of reason, human action, and intellectual inquiry. It is no surprise then that the linear time-image put forward by the now-weakened Catholic Church would be called into question in the New Age and would be found lacking. The philosophers of modernity saw it as an age defined primarily by the import it allotted to the future, by the extent to which it opens itself up to the horizon of expectation that the future casts and allows its present actions to be determined by its anticipation of what the future may hold.

In an age where the Christian faith that deeply infused Augustine's exploration of time was being increasingly dislodged and unsettled by skeptical inquiry, the notion of time as a stable concept that had a defined end of God's choosing and was seen as a progressive realization of redemption started to seem less tenable. As Agamben writes, "The modern concept of time is a secularization of rectilinear, irreversible Christian time, albeit sundered from any notion of end and emptied of any other meaning but that of a structured process of before and after",[30] whereas modern time retained the linear form bequeathed to it by Christian theology, it stripped it of its finalism. In the modern age, Christian time began to be secularized, which transformed the way in which it viewed the future. It contained a crucial new temporal relationship between the present and the future, which when stripped free of the religious frame of an imminent doomsday and therefore allowed for a more open conception of the future as a source of endless potential and progress that was expressed in the "conviction that the future has already begun: It is the epoch that lives for the future, that has opened itself up to the novelty of the future".[31] As David Carvounas writes, "Once past experience no longer tethered expectations, envisioning a future decidedly different from the past became possible. With the birth of modernity the past became undermined and the future valorized . . . The future . . . becomes in the modern age the focal point for a new organizing principle or transcendental anchor".[32] Not only was the future stripped by some of its religious implications; it was also given new meaning as the focal point of temporal experience and eagerly anticipated as a source of change and as a limitless horizon of expectation. Time, stripped of its eschatological meaning, was now "no more than a simple succession of now in terms of before and after".[33] In Agamben's terms, what was previously a rich record of the history of salvation had become "pure chronology" and, consequently, "a semblance of meaning [could] be saved only by introducing the idea . . . of a continuous, infinite progress".[34] Modernity had to give this new concept of temporality a fresh source of meaning, consonant with the rational and intellectual focus of its age. In Habermas's famous words, modernity as the "new age", with an empowered present and an anticipation of an extraordinary future therefore could no longer "borrow the criteria by which it takes its orientation from the models supplied by another epoch; it had to create its normativity out of itself".[35] The sites of meaning supplied by the Christian faith could no longer fit the dynamically new and the rational age, which refused to borrow the norms of a weakened past whose very institutions and traditions it was collapsing. In other words, "[t]he modern world needed a new source or a new kind of light to provide the present with a general sense of continuity, meaning and purpose".[36] It found this new site of meaning and purpose in the idea of progress and of man's constant improvement and development, in positing a utopian future where perpetual peace could someday be realized.

THE IDEA OF PROGRESS AND THE PROPER USE OF REASON: KANT'S UNIVERSAL HISTORY

Perhaps no philosopher was more deeply engaged in articulating this new source of meaning as Immanuel Kant, and it is his attempt to graft on to the linear concept of time inherited from the Christian faith, modern concepts of progress, development, purposiveness, and the regulative laws of reason that emerge as foundational to the modern canon. Kant's writings capture two impulses of modernity—the desire to break away from the antiquated traditions of the past and the need to refashion those normative concepts for the distinctively new age. Kant wants to divest linear time from the explicitly Christian meaning of time as the progressive realization of redemption as Augustine characterized it, as a time that begins with Creation and ends with the Day of Judgment, as distinct from the tranquility of Eternity. At the same time he also attempted to refashion morality and meaning for the new age, to revise the Augustinian concept of transcendence through faith with a transcendental metaphysic of regulative ideas governed by the proper use of reason. As Gordon Michalson writes in *Fallen Freedom*, Kant was faced with a dilemma symptomatic of his age and was caught in the tension between the developments of the new age and the displacements and loss of meaning it engendered—"this set of developments presented enormous difficulties for the thinker who wanted to embrace the new, scientifically based outlook while remaining, in some sense at least, 'religious'. In particular, there seemed to be no obvious way in this new situation to articulate a concept of transcendence that would do justice to the powerful new sense of human subjectivity and control while still retaining meaningful continuity with traditional language about God and divine action".[37]

The Kantian concepts of progress, the law-like model of nature and the categorical moral imperative are engendered by this twin impulse—to simultaneously break from the Christian tradition of revelation and to rewrite some of its tenets so that they could be secured by the postulates and hopes of reason instead. Kant's vast metaphysical project, set out in his three *Critiques*, of the proper use of pure reason, practical reason, and judgment, is outside the scope of this study. But several of the concepts he develops in those projects are also at the forefront in his *Political Writings*. In these essays, Kant is concerned with the Enlightenment consciousness, in articulating an idea of progress, in putting forward the notion of a purposive nature, and in establishing the primacy of reason in understanding. A close reading of these essays reveals the twin Kantian impulse to refashion temporal understanding for the modern age and to create a mode of thought that is transcendental, intellectualist, and skeptical of the sensory distractions of lived experience.

In *What Is Enlightenment?* Kant articulates this new age and its foundational concepts, defining it as a surfacing from the shadows of the received wisdom of the ages. In a key passage, he writes, "Enlightenment is man's emergence from self-incurred immaturity . . . This immaturity is *self-incurred* if its cause is not lack of understanding, but lack of resolution and courage to use it without the guidance of another. The motto of enlightenment is therefore: *Sapere aude!* Have courage to use your *own* understanding!"[38] Kant is acutely aware of the epochal significance of his age as the emergence from a superstitious world of signs and one infused with deep faith to one that is self-consciously regulated by the laws of reason. Enlightenment requires courage—to resist the edicts of the past, to question the received wisdom, to instead rely on one's individual faculties of understanding, and to resist the temptation to follow the whims and inclinations with the governance of reasons. Kant's sentiments simultaneously echo and depart from those expressed by Augustine. They are similarly suspicious of human inclination and passion and share an impulse to transcendence, but whereas Augustine sought guidance and direction in faith and confession, Kant looks for his revised solution in the language of moral imperative and the proper use of reason. For Kant, questions of religion are, on one hand, "the focal point of enlightenment" and the stripping away of its dogmas the clearest indication of "man's emergence from his self-incurred immaturity",[39] but, on the other hand, his need to locate a revised site of meaning for the new age is profound. As Michalson writes, "His difficulties are the difficulties of his historical setting rather than of philosophical argumentation. The awkwardness of his position is thus symptomatic of the fate of every religious liberal, whose lot it apparently is to live in a zone of discomfort, somewhere between modernity and tradition".[40]

Kant is aware of the significance of the moment he inhabits. It may be a time that is unsettled and not fully developed in its potential, but there are indications that the image of time, or the understanding of it, is changing. He writes, "If it is now asked whether we at present live in an *enlightened* age, the answer is: No, but we do live in an age of *enlightenment* . . . But we do have distinct indications that the way is now being cleared for them to work freely in this direction, and that obstacles to universal enlightenment, to man's emergence from self-incurred immaturity are gradually becoming fewer".[41] Kant's reflections on the Enlightenment and his concern with its displacement of traditional sites of meaning manifest itself in a revised understanding of temporality—time is no longer explicitly bounded by God and by the teachings of the Christian faith—but it invokes a future that is open, in which man can increasingly develop in maturity, in which obstacles to universal enlightenment and the peace and prosperity it can bring are decreasing, but one that is still governed by a moral imperative. In Kant's characterization, the future is something to be hopeful about, to anticipate, and to work toward, not the relentless tide of change that frustrated Augustine and made him yearn for the stillness of eternity.

Kant's reflections on time and history therefore represent a modern shift from those of Augustine even as he continues to revise him. For Kant, it is essential that the Christian faith be revised and reincorporated into the very framework of understanding and sites of meaning for the New Age. Kant wants to retain Augustine's faith in a different form—he wants to revise it for modernity and place it at the root of all of the new age's structuring principles. In his writings, this revision takes the form of the moral law, the categorical imperative, which cannot be proved or known but must be apodictically recognized. Kant's reflections on cosmopolitanism, on progress, and on development stem from the practical requirements of this apodictic recognition of the moral imperative. They are difficult to illuminate without grasping this essential formulation—it is impossible to grasp the Kantian image of time without recognizing that these powerful and modern concepts rest first and foremost on the moral imperative and are necessary postulates because of it. As a result, morality as law occupies a privileged and defining space at the center of the Kantian time-image, which brings him much closer to Augustine than he might first appear.

It may appear that the regulative ideals Kant identifies in universal history, nature, and providence in his essays, can be discerned empirically. For instance, Kant posits a notion of a future that is full of potential and can be shaped by individual understanding and reason, which initially appears to serve as a powerful challenge to the previously established concept of the future as having a defined end according to the Christian faith. However, such a shift in the understanding of futurity requires a concurrent and fundamental shift in the normative concepts that accompany such understanding. If God will no longer bring the future to a preordained end but it will instead continue indefinitely, a new set of ideas is required to give it meaning and salvage it from becoming a source of doubt and dread. In *Idea for a Universal History for a Cosmopolitan Purpose*, Kant puts forward such a set of regulative ideas that are distinct from those of the premodern age—namely, the idea of progress and its engine of development, a law-like and purposive model of nature, and the proper use of reason for understanding. At the beginning of the essay Kant links two powerful concepts of freedom and free will with the nascent idea of progress. He writes,

> Whatever conception of the freedom of the will one may form in terms of metaphysics, the will's manifestation in the world of phenomena, i.e. human actions, are determined in accordance with natural laws, as is every other event. History is concerned with giving an account of these phenomena, no matter how deeply concealed their causes may be, and it allows us to hope that, if it examines the free exercise of the human will *on a large scale,* it will be able to discover a regular progression among freely willed actions.[42]

According to Kant, it appears as though the will's actions in the world of experience, its exercise of freedom of action, is always determined by the laws of nature, and even though the precise mechanism of their operation might not be completely understood, a study of history, of human experience in time, will always reveal a "progression among freely willed actions". It suggests that such a regulative idea of constant progress, which imposes order on the newly liberated future and imbues it with hopeful anticipation, can be empirically validated, on the "large scale" of history, even if the vantage point from the present moment makes such validation difficult. Kant contends that if one looks at history, or the historical past, from the standpoint of the entire human species one can detect development over time. What appears in "individuals as confused and fortuitous may be recognized in the history of the entire species, as a steadily advancing but slow development of man's original capacities".[43]

Having advanced a theory of progress and slow development over time, Kant admits that it is difficult to find evidence of such a linear progression, if only because so little time has passed in the history of humankind and there are eons left to travel, so all examinations of development are conducted with a relatively small sample. He writes, "The real test is whether experience can discover anything to indicate a purposeful natural process of this kind. In my opinion, it can discover *a little*; for this cycle of events seems to take so long a time to complete, that the small part of it traversed by mankind up till now does not allow us to determine with certainty the shape of the cycle".[44] The critical question to ask at this juncture is, Why is it necessary to assume that the future is ordered by a regulative idea of progress and linear development? Why must the idea of purposiveness, as the motor of the engine of development, become a necessary philosophical postulate? The answer lies in the foundational role that the moral imperative plays in Kant's thoughts on this subject. For him, morality takes the form of law that is binding on everyone and can be apodictically recognized by all and it is therefore not something he can empirically prove or know as such. As William Connolly has written, "Kant treats this categorical judgment about the very form of morality itself as a fundamental recognition always already achieved by every ordinary person who has not been confounded by individual perversity or thrown astray by the dictates of an underdeveloped civilization".[45] The categorical imperative therefore operates at the center of Kant's philosophy as a fact of "fundamental recognition" that cannot be contested. As such, it is this apodictic recognition of the moral law that necessitates the postulates of progress, development and purposiveness that give rise to Kant's image of time. As Connolly writes, Kant's cosmopolitanism "gains a grip in Kant's moral imagination only because it flows from a transcendental imperative to act *as if* the world were filled with a *providential direction* that stretches beyond the reach of human agency. Kant's moral transcendentalism enables cosmopolitanism to function as a regulative ideal in his time".[46]

In his essay on *Theory and Practice*, Kant makes this connection between the apodictic recognition of morality and the consequent and necessary assumption of progress clear. He writes:

> I may thus be permitted to assume that, since the human race is constantly progressing in cultural matters (in keeping with its natural purpose), it is also engaged in progressive improvement in relation to the moral end of its existence. This progress may at times be *interrupted* but never *broken off* . . . I base my argument upon my inborn duty of influencing posterity in such a way that it will make constant progress

> (and I must assume that progress is possible), and that this duty may be rightfully handed down from one member of the series to the next.[47]

Here Kant suggests that it is not a matter of the empirical discernment of progress in human history, but that such progress is necessary in "relation to the moral end of its existence". For the human race to progress at all times is an "inborn duty". We must therefore "assume that progress is possible" if we are to act in accordance with the moral law and follow our duty toward this moral end. It suggests that progress itself is a necessary postulate—it is not simply empirically evident from the vast sweep of history but something that must be posited in order to realize the moral end of existence. In Connolly's words, Kant demands that "despite what actual history might teach, that we act *as if* nature and history are intelligible. It suggests, further that we act *as if* there is a guiding thread in history that pulls us, beyond the reach of effective human agency".[48] As such, the idea of cosmopolitanism, the idea of progress, the idea of providence in nature and history, "all flow from the same slender reed of apodictic recognition at the base of Kantian morality".[49]

Thus, the philosopher must "attempt to discover a purpose of nature behind this senseless course of human events".[50] Indeed, for Kant, this is a significant philosophical effort to impose order on the course of world history, akin to the scientific theories of the Enlightenment. He compares this attempt, to theorize a guiding principle and to then discern its pattern to Kepler's and Newton's successful regulating of the eccentricities of the planets and forces of motion by definite laws. It is a profound attempt at structuring lived experience and quieting its disruptive moments. Thus, for Kant, for whom the fundamental recognition of morality as law operates at the center of his philosophy, it is critical to posit a law-like model of nature where natural laws regulate and determine will in the world of phenomena, in order to establish a purposiveness, an idea of progress and development. Kantian morality therefore requires that Nature appear to be governed by law-like regularity and efficient causality.

Therefore, in what appears at first to be a direct contrast to Augustine's invocation of God and faith to provide salvation from the constant flux of human time, Kant actually ends up performing a similar quieting move by turning to Nature to regulate an otherwise chaotic and relentlessly impending future and assigning to it a concept of development that can undergird and order it. For Kant, the course of human history is that of an overall progression, perhaps even a plan, but not one explicitly laid out and closed off by God. He writes that "[i]ndividuals, men and even entire nations little imagine that, while they are pursuing their own ends, each in his own way, and often in opposition to others, they are unwittingly guided in their advance along a course intended by nature".[51] Time, in all its immensity, unfurls according to nature's plan, moving towards an ideal end of human development. Such a regulative concept is essential to the Kantian understanding of time, where the future emerges in a key role and contrasts with the explicitly Christian understanding of time coming to an end at God's choosing. Kant makes this point clear when he writes of this assumption of purposiveness: "But if we assume a plan of nature, we have grounds for greater hopes. For such a plan opens up the comforting prospect of a future in which we are shown from afar how the human race eventually works its way upward to a situation in which all germs implanted by nature can be developed fully, and in which man's destiny can be fulfilled here on earth".[52] Assuming nature has a plan that it executes for man's development then allows the future to appear welcoming and desirable, as a horizon of limitless expectation. This key assumption of the purposiveness of nature not only provides for the possibility of a "comforting" future, but it is also a necessary postulate for the Kantian moral law. We must act *as if* there is a plan of nature, we must "assume a plan of nature", which will give us "grounds for greater hopes" and allow us to realize the moral end of existence.

This apparent secularization of Christian time, where nature seemingly supplants God as the source of temporal order therefore masks the much deeper moral imperative that is operative in Kant's writing. Kant conflates nature with providence as a useful tool for understanding as he writes, "Such a *justification* of nature—or rather perhaps of *providence*—is no mean motive for adopting a particular point of view in considering the world".[53] Kant is somewhat wary of providence, but he is acutely aware that the use of nature's purposiveness plays a similar role in temporal understanding. He is convinced that such a regulative idea of nature and the consequent philosophical attempt to write history in conformity with it is essential, even if it is cannot be proved to be true, as any attempt at order is preferred to the ordinary chaos of human actions. In a revealing passage, he writes, "Yet if it may be assumed that nature does not work without a plan and purposeful end, even amidst the arbitrary play of human freedom, this idea may nevertheless prove useful".[54]

Kant appears to be trying to rework Augustine's faith in an omnipotent God with the structuring heft of a purposive and law-like model nature, in which the idea of progress functions as a regulative idea for the understanding of history. However, although he appears to be supplanting the Christian faith that is central to Augustine with a belief in progress, Kant does not overturn the linear model of time; instead, he further reifies it. Time, for Kant, still takes the shape of a straight line, moving incessantly from the past to a fleeting present to a future that is open and full of potential and is the transcendental anchor for this time-image. Kant frees linear time from its defined Christian end of the Day of Judgment, but in redefining it for the modern age, he only further strengthens its teleological vector. The telos of this modern time is no longer the promised moment of salvation but the development of the human species as a whole toward an idealized end as laid out by a law-like model of nature in accordance with the moral law. He writes, "All the natural capacities of a creature are destined sooner or later to be developed completely and in conformity with their end".[55] Taken as a whole, as a species, over generations, man can therefore develop and reach nature's intention. Kant makes this explicit when he writes,

> It will require a long, perhaps incalculable series of generations, each passing its enlightenment to the next, before the germs implanted by nature in our species can be developed to that degree which corresponds to nature's original intention. And the point of time at which this degree of development is reached must be the goal of man's aspirations (at least as an idea in his mind), or else his natural capacities would necessarily appear by and large to be purposeless and wasted.[56]

In Kant's terms, the original destiny of man, his teleological end, lies in such progress, even if this development is only an "idea in his mind" as it regulates his natural capacities and the actions of his will, which might otherwise be wasted without such direction.[57] What Kant himself terms this "teleological theory of nature" is therefore essential to understanding human history and temporality, "for if we abandon this basic principle, we are faced not with a law-governed nature, but with an aimless, random process, and the dismal reign of chance replaces the guiding principle of reason".[58] The trade-off for Kant is clear—it is a choice between a throw of the dice of chance that results in random chaos or a time-image that is governed and undergirded by the laws of nature.

Critical to this choice for Kant, is the development of an image of thought to accompany this orderly, linear, and teleological concept of time. The guiding principle of reason and its proper use therefore emerges as central in the transcendental philosophy Kant develops. For Kant, it is reason that enables progress and development, and the proper use of reason allows him to posit an unbounded future that casts a horizon of expectation that we can fill with the space of our experience. He writes, "Reason, in a creature, is a faculty which enables that creature to extend far beyond the limits of natural instinct the rules and intentions it follows in using its various powers, and the range of its projects is unbounded".[59] This transcendental image of thought echoes Augustine in the emphasis it places on ordering the world of lived experience. While Augustine chose to privilege the tranquility of eternity over the vertiginous experience of time in everyday life when he contended that faith in God allowed man to exceed his limitations and the carnal temptations of the life of the flesh, Kant puts forward a secular rendering of the linear time image that strips it off its religious significance and end of days finalism but keeps its quieting imperative intact. Similar to Augustine, Kant diminishes the role of instinct and of the body, suggesting that it is only that which is produced by reason that can lead to true happiness, writing, "nature has willed that man should produce entirely by his own initiative everything which goes beyond the mechanical ordering of his animal existence, and that he should not partake of any other happiness or perfection that that which he has procured for himself without instinct and by his own reason".[60] Kant therefore puts forward an image of thought to accompany this modern model of linear time that privileges reason throughout, perhaps elevating it to the structuring imperatives of faith and echoes Augustine in the transcendental impulse to impose order on the distractions of lived experience in his dismissal of instinct and sentiment.

Kant's modernization of temporal understanding also seeks to refashion morality for the new age. Kant's debt to Augustine becomes increasingly clear as he seeks to carve out a place for God and morality within his philosophy, without resorting to the expressions of revelational faith that the Enlightenment disavowed. He writes in his reflections on *Perpetual Peace* that "if we consider its purposive function within the world's development, whereby it appears as the underlying wisdom of a higher cause, showing the way towards to objective goal of the human race and predetermining the world's evolution, we call it *providence*".[61] This introduction of providence is couched in the rational language of purposive function. Even though we cannot observe the operation of providence in nature and deduce its existence rationally from it, "as with all relations between the form of things and their ultimate purpose, we can and must *supply it mentally* in order to conceive of its possibility . . . Its relationship to and conformity with the end which reason prescribes to us (i.e. the end of morality) can only be conceived of as an idea".[62] Indeed, it is only the fear of overstepping the limits of reason that motivates Kant to speak of such regulation as that of nature and not of providence. Kant writes, "It is more in keeping with the limitations of human reason to speak of *nature* and not of *providence,* for reason, in dealing with cause and effect relationships, must keep within the bounds of possible experience. *Modesty* forbids us to speak of providence as something we can recognize".[63] It is only modesty, or a certain rational humility, that prevents Kant from introducing God into his metaphysical framework, even as he establishes morality as law at its core that seems to require a tacit recognition of providence, within the limits of reason. In that he is much closer to Augustine than he first seems.

Kant's debt to Augustine's reflections become even clearer when he introduces the language of the categorical imperative and the adherence to moral law as essential to man's development and his ability to harness the future. The idealized goal of perpetual peace is guaranteed by nature, and it "makes it our duty to work our way towards this goal, which is more than empty chimera".[64] The language of imperative and duty, of conforming to an *a priori* moral law is deeply evident in his essay on *Theory and Practice*. There Kant states that "this concept of duty does not need to be based on any particular end, but rather itself occasions a new end for the human will, that of striving with all one's power towards the universal happiness of the whole world, combined with and in keeping with the purest morality".[65] This concept of duty that is now deeply embedded in the regulative idea of progress is the primary impetus toward an ideal future as it compels man to strive harder to attain this happy end. Not only is progress a regulative idea that gives the future meaning; it is also *a duty of the human will to develop according to the moral law.* Indeed, the idea of progress itself is a necessary postulate of the moral law. As Howard Caygill has argued, Kant seeks to develop a moral philosophy on the basis of this autonomous principle of self-legislation, such that "the idea of freedom as autonomy is thus uncovered as the ground for the categorical imperative, the necessary presupposition accorded by us to ourselves and other rational beings in so far as they possess a will".[66]

In making the categorical imperative, or the recognition of the moral law one gives oneself, central to his philosophical project and the attendant image of time, Kant reveals his indebtedness to Augustine. Indeed, faith in a higher being becomes a necessary postulate for the Kantian temporal understanding. Kant writes, "Since the attainment of this good lies within our power . . . it elicits from our reason a faith, *for practical purposes*, in a moral being who governs the world, and in a future existence . . . it merely ensures that this concept acquires an object in the shape of an ideal of pure reason".[67] The moral imperative and faith in a higher being work as a practical necessity for positing such a future. They do so because of the practical "necessity of assuming as the ultimate end of all things a *highest good* on earth" and especially because "there can be no *will* without an end in view".[68] The free will cannot exist if there is no ideal, necessary future end to which it must continuously strive. To let Kant finally speak in the voice of Augustine, "No idea can so greatly elevate the human mind and inspire it with such enthusiasm as that of a pure moral conviction . . . the fact that man is aware the he can do this just because he ought to discloses within him an ample store of divine capabilities and inspires him, so to speak, with a holy awe at the greatness and sublimity of his true vocation".[69] The belief in God is necessary for Kant because in the absence of such a faith and the regulation, it provides through the categorical moral imperative the world would be in chaos. As Michalson writes, "As a good Kantian, then, I believe in God because I cannot conceive that the universe is absurd and asymmetrical . . . Indeed, it is just this trust in the rationality of the universe that gives morality its point. In *this* universe, moral endeavor is ultimately meaningful and makes a positive difference. Belief in God and confidence in morality grow out of the same metaphysical soil".[70]

Kant's affinity to Augustine is further revealed in his writings on radical evil. Kant's philosophical project and its attendant image of time hinge on the recognition of the moral law and on acting in accordance with its imperative, which echoes Augustine's faith in God and the necessity of acting in harmony with his divine plan. In addition, Kant also locates the idea of evil in the will, in a manner that parallels Augustine. Kant cannot conceive of evil as occurring in nature as the universe is ordered and harmonious, not chaotic and unruly and therefore he locates evil in the will itself. In *Religion within the Limits of Reason* Kant writes,

> There is in man a natural propensity to evil; and since this very propensity must in the end be sought in a will which is free, and can therefore

> be imputed, it is morally evil. This evil is *radical*, because it corrupts the ground of all maxims; it is, moreover, as a natural propensity *inextirpable* by human powers . . . yet at the same time it must be possible to *overcome* it, since it is found in man, a being whose actions are free.[71]

His account of the free will as being the ground of radical evil, which therefore corrupts its maxim making and makes it act in ways that do not accord with the moral imperative is closely connected to Augustine's concept of evil as being lodged in the division of will which can only be healed by grace. In Michalson's terms, it is almost as though each of us "carries a malevolent stowaway that could come to life at any moment without cause or explanation"[72]. According to Michalson, Kant's "basic definition of moral evil turns on the threefold idea that: sensuous incentives gain the upper hand in the formation of maxims, sensuousness is naturally given, and we are therefore not to be blamed for ways in which sensuous incentives affect us, and perhaps even over what sensuous inclinations we have"[73]. In his discussion of the problem of evil, which he locates within the will itself, Kant betrays a similar aversion to the passions and desires of the body that Augustine does. Michalson writes, "For although he is careful not to blame moral evil on the sheer fact of our bodiliness, it is still the case that moral evil arises out of the competition for control of character between sensuous and rational incentives: rationality would have no competition if it were not embodied".[74] In his account of radical evil Kant reveals his indebtedness to Augustine, such that even as he secularizes much of the latter's conceptual framework for the new age with a belief in progress and a linear account of time, in the final analysis, he reintroduces the problem of evil into the world and initiates a similar turning away from the body and its sensuous desires. For Kant, the act of defining temporal understanding for the modern age therefore requires a simultaneous act of detachment from and entrenchment of the Augustinian understanding of time. While he frees linear time from its preordained Christian end, the image of thought that he posits to give it ontological meaning betrays the same transcendental impulse as that of Augustine. Such a future requires regulative ideas of progress and a law-like model of nature, a privileging of the faculty of reason, and, for its chief motivation, a categorical imperative of accordance with the moral law, where faith and belief in a higher moral being is reintroduced as a necessary postulate for human freedom.

THE TRANQUILITY OF "OUR TIME"

Tracing the emergence of the dominant concept of "our time", as Sheldon Wolin has called it, which has provided the common narrative for the theorizing of the political, reveals an image of time that is linear in form and is wrapped in a transcendental image of thought. Linear time emerged as a dramatic challenge to ancient notions of cyclical time and was a product of the Christian faith that structured it throughout and is bounded at its beginning and end by Genesis and the Day of Judgment, respectively. Time, for Augustine, was created by an omnipotent God who would bring it to a preordained end. It was seen as the progressive realization of redemption, a dizzying confusion, which must be transcended by a deep faith in God, a purging of the temptations of the flesh, and a yearning for the stillness of eternity. In the age of modernity, amid the Enlightenment, Kant sought to revise this linear concept of time and to open the future as a source of unlimited potential. In doing so, Kant reconfigured the future as a horizon of expectation, where an idealized end could be achieved. "Our time" remained linear in form, but it now had the future as its anchor, and was regulated by an idea of progress, of a purposive nature and the proper use of reason. In modernity, "our time" has shaken off its preordained Christian end on the Day of Judgment, but it has further entrenched and revised its impulse to transcendence—it is now anchored in a categorical moral imperative and faith in a higher being, in a turning away from the body and its sensuous desires and an accompanying privileging of reason, in the lodging of evil in free will itself, and in a whisper of providence throughout. This exploration of "our time", which makes sense of another who operates outside of it, reveals within itself a culturally constructed set of philosophical concepts that owes much to its Christian roots and the transcendental image of thought that sustains it and gives it purpose. Complicating such an image of time in order to confront the challenges of theorizing in a diverse world of multiplicity, requires that a challenge be posed to the dominant image of thought engendered by "our time", by taking into account the role of memory and the density of perception. The following chapter engages in such an attempt—to explore the conceptual resources that can constitute a more complex account of time and the challenges to the transcendental image of thought that such ideas might foster.

NOTES

1. Georgio Agamben, "Time and History—Critique of the Instant and the Continuum", in *Infancy and History: Essays on the Destruction of Experience*, trans. L. Heron (New York: Verso Press, 1993), p. 96.
2. David Carvounas, *Diverging Time: The Politics of Modernity in Kant, Hegel and Marx* (Oxford: Lexington Books, 2002), p. 6.
3. Ibid., p. 7.
4. J. G. A. Pocock, *The Machiavellian Moment: Florentine Political Thought and the Atlantic Republican Tradition* (Princeton, NJ: Princeton University Press 1975), p. 6.
5. Ibid., p. 7.
6. Agamben, *Time and History*, p. 94.
7. Pocock, *The Machiavellian Moment*, p. 31.
8. Carvounas, *Diverging Time*, p. 7.
9. Agamben, *Time and History*, p. 95.

10. Augustine, *The Confessions*, trans. M. Boulding (New York: Vintage Spiritual Classics, 1997), p. 37.
11. Ibid., p. 43.
12. Ibid., p. 250.
13. Ibid., p. 254.
14. Ibid., p. 252.
15. Ibid., pp. 255–256.
16. Ibid., p. 259.
17. Ibid., p. 267.
18. Ibid., p. 271.
19. Agamben, *Time and History*, p. 95.
20. Augustine, *The Confessions*, p. 271.
21. Ibid., p. 138.
22. Ibid., p. 120.
23. Ibid., p. 120.
24. Ibid., p. 120.
25. Hans Blumenberg, *The Legitimacy of the Modern Age*, trans. Robert E. Wallace (Cambridge, MA: MIT Press, 1983), p. 133.
26. Augustine, *The Confessions*, p. 163.
27. Augustine, *City of God*, trans. Henry Bettenson (London: Penguin Books, 1972), p. 503
28. Ibid., p. 503.
29. Jürgen Habermas, "Modernity: An Unfinished Project", in *Habermas and the Unfinished Project of Modernity: Critical Essays on the Philosophical Discourse of Modernity*, ed. M. Passerin d'Entrèves and S. Benhabib (Cambridge: MIT Press, 1996), p. 45.
30. Agamben, *Time and History*, p. 96.
31. Jürgen Habermas, *The Philosophical Discourse of Modernity*, trans. Frederick G. Lawrence (Cambridge, MA: MIT Press, 1995), p. 5.
32. Carvounas, *Diverging Time*, p. 12.
33. Agamben, *Time and History*, p. 97.
34. Ibid.
35. Habermas, *The Philosophical Discourse of Modernity*, p. 7.
36. Carvounas, *Diverging Time*, p. 12.
37. Gordon E. Michalson, *Fallen Freedom* (Cambridge: Cambridge University Press, 1990), p. 2.
38. Immanuel Kant, "An Answer to the Question: 'What Is Enlightenment?'", in *Political Writings*, ed. H. Reiss and trans. H. B. Nisbet (Cambridge: Cambridge University Press, 1991), p. 54.
39. Kant, "What Is Enlightenment?", p. 59.
40. Michalson, *Fallen Freedom*, p. xi.
41. Kant, "What Is Enlightenment?", p. 58.
42. Immanuel Kant, "Idea for a Universal History with a Cosmopolitan Purpose", in *Political Writings*, ed. H. Reiss and trans. H. B. Nisbet (Cambridge: Cambridge University Press, 1991), p. 41.
43. Ibid., p. 41.
44. Ibid., p. 50.
45. William E. Connolly, *Neuropolitics: Thinking, Culture, Speed* (Minneapolis: University of Minnesota Press, 2002), p. 180.
46. Ibid., pp. 179–180.
47. Immanuel Kant, "On the Common Saying: 'This May be True in Theory, but It Does not Apply in Practice'", in *Political Writings*, ed. H. Reiss and trans. H. B. Nisbet (Cambridge: Cambridge University Press, 1991), pp. 88–89.
48. Connolly, *Neuropolitics*, p. 182.

49. Ibid., p. 182.
50. Kant, "Idea for a Universal History with a Cosmopolitan Purpose", p. 42.
51. Ibid., p. 41.
52. Ibid., pp. 52–53.
53. Ibid., p. 53.
54. Ibid., p. 52.
55. Ibid., p. 42
56. Ibid., p. 43.
57. Kant, "What Is Enlightenment?", p. 57.
58. Kant, "Idea for a Universal History with a Cosmopolitan Purpose", p. 42.
59. Ibid.
60. Ibid., p. 43.
61. Immanuel Kant, "Perpetual Peace", in *Political Writings*, ed. H. Reiss and trans. H.B. Nisbet (Cambridge: Cambridge University Press, 1991), p. 108.
62. Ibid., p. 109.
63. Ibid.
64. Ibid., p. 114.
65. Kant, "On the Common Saying", p. 65.
66. Howard Caygill, *A Kant Dictionary* (Cambridge, MA: Blackwell Publishers, 1995), p. 101.
67. Kant, "On the Common Saying", p. 65.
68. Ibid., p. 65 (extended footnote).
69. Ibid., p. 71.
70. Michalson, *Fallen Freedom*, p. 25.
71. Immanuel Kant, *Religion within the Limits of Reason* (New York: Harper & Row Publishers, 1990), p. 32.
72. Michalson, *Fallen Freedom*, p. x.
73. Ibid., p. 47.
74. Ibid., p. 69.

3 Complex Time, Memory, and the Ethics of Affirmation

The image of time that structures and shapes much of Western philosophy is so familiar that it can simply be termed "our time". It envisions a sequence of moments moving effortlessly, irreversibly, and relentlessly from a past that disappears to a present that slips away in a moment, to a future that is close to our grasp. "Our time" is readily represented in spatial terms—it takes the form of a straight line, and its three stages of past, present, and future can be sequentially mapped as points along that line. In mathematical terms, the line takes the form of a vector—an arrow signaling the forward motion from the past to the present to the future. In philosophical terms, this spatial image is sustained by teleology—whether it carries the redemptive Christian hope of salvation or the Enlightenment's faith in progress. But what of the time that is less familiar in philosophy but most familiar in everyday life? What of the past that never leaves us—the fond memories of a grandfather telling stories that we can recall with vivid clarity and the song on the radio that whisks us back to high school in a cloud of recognition and embarrassment? What of the trauma of a car wreck that is never quite forgotten, that grips our minds when driving at night on a rain-slicked highway? What of those pregnant moments of anxious anticipation when five minutes feel like an hour and of moments of intense concentration when two hours flit by in what feels like ten minutes? How can such complex experiences of time as out of joint be captured by the thin line that relegates the past to a mere point along a sequence and not as ever-present memory?

This familiar time of lived experience has been captured in vivid detail in Hamlet's experience of time as out of joint. The complex interplay between encounter and memory in perception and its attendant dislodging of sequential time is illustrated powerfully in cinema too, from the whirlpool of terror and confusion that is Hitchcock's *Vertigo* to the fleeting and feverish memories of Christopher Nolan's *Memento*, to name a few. In the realm of philosophy, however, complex accounts of time, while scattered throughout the history of the discipline, have less often been explored. Instead, as Sheldon Wolin has observed, it is the linear, sequential, and teleological "our time" that serves as the foundation for common understandings of identity, a shared framework of ethics, and our concepts of the political. The aim of this chapter is to displace this narrative of "our time", to dislodge its neat sequence with accounts that more accurately capture the time-image of lived experience and to ask what consequences the adoption of such an alternative view may have for philosophical understanding. A complex and nonsequential account of time is one where the past, present, and future coexist, sometimes merging fluidly and sometimes awkwardly with one another; where memory and affect cause the past to rub up against an expanding present; and where the anxieties of the creative future also shape and shove the now. There is not one rigid and linear sequence of past, present, future, but one where the order is always open to disruption. It is a narrative of time that is more representative of our experience of time, where we draw constantly on memory, when the past of recollection rushes into the present, or when we feel the anxiety of future demands impinging on the present. Complex time is the implicit and intuitive time underlying experience, before it emerges as the current that is neatly quantified by the precepts of the linear time-image.

Complex accounts of time have subsisted in fragmentary form in much of classical and contemporary thought. While they may vary in their emphasis, at their core, they share an attentiveness to memory, a sense of the new, a strong element of affirmation, and an awareness of creative potential instead of a concept of sufficient causality. One such account of nonlinear time can be found in Stoicism. The central feature of the Stoic conception is the *cairos*—the "abrupt and sudden conjunction where decision grasps opportunity and life is fulfilled in the moment".[1] In the *cairos*, both past and future can coexist in one powerful, innovative and authentic present. They are no longer sequentially distinct and do not obey a linear form. Even in this ancient form, the experience of complex time is not structured and disciplined by an external structure. There are various other accounts, in Gnostic texts; in novels, plays, and modern poetry; and certainly in existentialist and phenomenological thought. This chapter engages with two philosophers who explored in depth the complex intermingling of time, memory, and perception: Henri Bergson and Friedrich Nietzsche. Bergson lays out a complex image of time, highlighting the presence of the past and the role of memory by freeing time from its spatial representations. Nietzsche layers a profound sense of ethical and political urgency into the conversation by exploring what temporal understanding means for lived experience. Finally, this chapter turns to the temporal riddle of Eternal Return, which forms the cornerstone of Nietzsche's philosophy, to extricate from it an ethic of affirmation that embraces life as replete with difference, in a world that is perpetually becoming.

TIME AND MEMORY: THE BERGSONIAN TIME-IMAGE

What might be the characteristics of a complex account of time that is true to the lived experience of time as out of joint? In his seminal *Matter and Memory*, the French philosopher Henri Bergson pursues such a time-image.

Bergson's account of duration seeks to free time from its sublimation to spatial representations, allowing it to emerge instead as pure becoming, or alteration with continuity. With its layered account of the continual presence of the past and the anticipated future, it serves as a powerful contrast to the dominant notion of time as simple forward motion. In contrast to the transcendental image of thought that has formed the lynchpin for the linear concept of time, Bergson seeks to conceptualize duration from the lived experience of time. Intuition, not legislative reason, is the method of his philosophy. Intuition is not a mere feeling or sense of inspiration, but a rigorous method that allows one to examine experience. It is a method of division that helps us distinguish differences in degree from differences in kind and consequently to dispel the illusions that emerge from confusing one with the other. For Bergson, the aim is to construct metaphysics from the perception and recollection of lived experience. Bergson's philosophy is precisely not of a transcendental nature; it is instead an attempt to discern philosophical truths from experience itself. As one of his key interlocutors, Gilles Deleuze, writes, in Bergsonian philosophy, "we go beyond experience, toward the conditions of experience (but these are not, in the Kantian manner, the conditions of all possible experience: They are the conditions of real experience".[2] In Bergson's terms, the Kantian critique "succeeded only by limiting the range and value of our senses and of our understanding";[3] Bergsonism, instead, privileges common sense.

According to Deleuze, duration, for Bergson, is memory, consciousness, and freedom—it is the essential concept of his philosophy and his supplanting of space with duration is his greatest philosophical contribution. In *Matter and Memory*, Bergson provides a compelling metaphysical account of duration as complex time, emerging from perception by intuition. He removes from time the disciplining restraints of forward motion, of endless progress, and unshackles it from its strictly linear spatial representation. Instead he allows it to emerge as duration, which he defines as a complex flow of pure becoming. Duration contains within it a past that is always present, that gnaws at the future, a past that we access through leaps into its layers to extricate that which recollection calls up. Bergson's theory of duration is therefore deeply wrapped up in his theory of memory. As Bergson writes, "Memory, inseparable in practice from perception, imports the past into the present, contracts into a single intuition many moments of duration".[4] The whole of the past therefore conditions our present state without necessarily determining it. All of our past experiences are always available to us at any present moment and impact it, such as when we are walking down the street and have a vague memory of a conversation that took place on such a walk a few days ago, and if we choose to focus on it, the conversation from the past is brought into our very present, as we remember specific snippets of dialog, the heat of the pavement, and the smell of the grass. These layers of memory continue to hover around the edges of the present and color and shape it, often without being explicitly recollected into the present. They form a fog of condensation that influences the present even when they are not explicitly called up to act on it.

Thus, while the past is always present, either inert and gently influencing the present, or explicitly conditioning our present state when we call it up by recollection, the present is that which is continuously being made—it is the indivisible limit which divides the past from the future. In lived consciousness it appears that the present consists only in the immediate past—it slips away before it is congealed in our minds as such; we recognize the present only when it has passed. However fleeting, each instant always contains within it a multitude of remembered elements, which are contracted into recollection. Memory and perception are inextricably bound up with each other, such that "in truth, every perception is already memory. *Practically, we perceive only the past*, the pure present being the invisible progress of the past gnawing into the future".[5] Duration therefore does not conform to the spatially determined coordinates of linear time. Indeed, it cannot be spatially represented, it is a stream of becoming where the whole of the past coexists in protraction of the present and is selectively contracted into it according to the dictates of perception and the needs and utility of life. The everyday moments of our lives trigger our memories that envelop us in the out-of-joint experience of time. In Bergson's words, "it is from the present that the appeal to which memory responds comes, and it is from the sensory-motor elements of present action that memory borrows the warmth which gives it life".[6] The self moves between the various layers of the past and the present, implicit and inert, or explicit and conditioning; moving through different planes and the meshes of necessity, it exists in its temporal movement. Memory thus enhances our consciousness—it is by the leap into the past and the contraction of memory that "a nebulous mass, seen through more and more powerful telescopes, resolves itself into a greater number of stars".[7] The past therefore coexists with the present, and our memories are often imbricated into perception of the present, so much so that they condition and color it. The boundary between memory and perception hardly exists. Rather, it is a constant intermingling, a shoving and shaping, a coloring of the everyday with the hues of memory that in turn helps to shape perception as well. Bergson writes,

> We assert, at the outset, that if there be memory, that is, the survival of past images, these images must constantly mingle with our perception of the present and may even take its place. For if they have survived it is with a view to utility; at every moment they complete our present experience, enriching it with experience already acquired, and as the latter is ever increasing, it must end by covering up and submerging the former. It is indisputable that the basis of the real, and so to speak instantaneous, intuition, on which our perception of the external world is developed, is a small matter compared with all that memory adds to it.[8]

The images of the past, the hazy snapshots of memory, are therefore always present, and we can call them up whenever we need them to enhance our understanding of an event or situation with the benefit of past experience. The past, for Bergson, is not simply a moment in time that slips quietly away into the ether of time. Instead, the past is ever present and a rich trove that constantly conditions and enhances experience. Although time is irreversible and the past can never be experienced in exactly the same manner in the future, we can leap into it and recall parts of the experience. Bergson writes, "Whenever we are trying to recover a recollection, to call up some period of our history, we become conscious of an act *sui generis* by which we detach ourselves from the present in order to replace ourselves, first, in the past in general, then, in a certain region of the past—a work of adjustment, something like the focusing of a camera".[9]

The past image, though, is virtual before parts of it take the form of recollection—"little by little it comes into view like a condensing cloud", it passes into the actual and tends to imitate perception—"but it remains attached to the past by the deepest roots, and if, once realized, it did not retain something of its original virtuality . . . we should never know it for a memory".[10] When we listen to a live recording of a concert we attended, our experience of the song—the soaring vocals, the wailing guitar, the pounding bass line, the roar of the crowd—is difficult to separate from our experience of being squeezed in a sweaty throng of new friends, arms raised in a V, and singing at the top of our lungs. The memory pushes its way so insistently into the present and colors and enhances every note of the song so intensely that if we didn't look around us and take note of the fact that we are on a train on our morning commute, it would feel as though we were almost there. It is precisely because the memory retains its original character and remains attached to the past with its deep roots that we recognize it as such.

Thus, there is a key distinction between the present and the past and that is one of action and of inaction. While memory can recall the past in the present, it cannot make it act in the present. The past is inert, but it conditions the present which can act—"The past is only idea, the present is *idea-motor* . . . the past is essentially *that which acts no longer*", and it is therefore essential to make "a real distinction between it and the present, i.e., *that which is acting*".[11] The present, for Bergson, is the locus of action. He writes that "there is much more between the past and the present than a mere difference of degree. My present is that which interests me, which lives for me, and in a word, that which summons me to action; in contrast my past is essentially powerless".[12] But the always-present past is not completely powerless—as we have seen, it contains memories that enhance and condition our present. In a dual reaction, the past conditions the present, and the present conditions the past. The past itself may be unable to act, but it works on the present and allows it to act. And the present itself calls up the memory and actualizes it and helps to give it form. As Bergson puts it, "it is from the present that the appeal to which memory responds comes, and it is from the sensori-motor elements of present action that a memory borrows the warmth which gives it life".[13]

The time-image for Bergson is that of a stream of becoming, it is a concept of time in which the familiar tenses of 'our time' mingle with each other in lived experience. He captures this nonspatial concept of time deftly when he writes,

> What I call "my present" has one foot in my past and another in my future. In my past, first, because "the moment in which I am speaking is already far from me"; in my future, next, because this moment is impending over my future: it is to the future that I am tending . . . the psychical state, then, that I call "my present", must be both a perception of the immediate past and a determination of the immediate future.[14]

The present moment is therefore not simply one more point along the vector of time; it is also a moment that is pregnant with the past and that anticipates and therefore contains the future—it perceives both that which preceded it and that which is to come. In Bergson's words, the present "consists in the consciousness I have of my body . . . thus it, indeed, represents the actual state of my becoming, that part of my duration which is in a process of growth".[15] The Bergsonian time-image, or duration, is a flow of becoming and the present represents a moment of its actualization. Duration, therefore, is not composed of separated and inactive states; it contains a past that can condition and be actualized in the present, a present that acts and becomes, and a future that is anticipated in the moment. It is not that time is not irreversible: it is that the relation of the past to the present to the future is much more complex and rich than is accounted for in teleological accounts of linear time.

Why then is it so difficult to come to terms with the out of joint experience of duration? Gilles Deleuze writes that we "have great difficulty in understanding a survival of the past in itself because we believe the past is no longer, that it has ceased to be".[16] Bergson insists that part of the problem is the trouble we have in conceiving a past that is present, in asking where in the cerebral substance such large quantities of it could be stored, in confusing the matter of brain cells with memory, which are different in kind. As such, we break up duration into discrete parts and mistakenly assign to the flow a calibrated sequentiality of instants. He writes,

> The survival of the past *per se* forces itself upon philosophers, then, under one form or another; the difficulty that we have in conceiving it comes simply from the fact that we extend to the series of memories, in time, that obligation of *containing* and *being contained* which applies only to the collection of bodies instantaneously perceived in space. The fundamental illusion consists in transferring to duration itself, in its continuous flow, the form of the instantaneous sections which we make in it.[17]

Such representation belongs to bodies in space, but not to time. To force time into a quantified sequential flow is a key error; it mischaracterizes duration as we experience it in life. It imposes on the flow of becoming a strict order that becomes almost impossible to sustain once we broach the subject of memory and the continued existence of the past.

The Bergsonian time-image therefore complicates the linear and teleological accounts of time, by debunking the idea of sequentiality that sustains it. If time were indeed sequential, there would have to be a gap or a marker between a moment that is present and a moment that is past, but is there such a boundary? As Deleuze writes, "If the past had to wait in order to be no longer, if it was not immediately and now that it has passed, 'past in general', it could never become what it is, it would never be *that* past".[18] The Bergsonian time-image, with its account of memory, complicates "our time" by displacing its neat sequentiality. In Bergson's account, "the past and present do not denote two successive moments, but two elements which coexist. One is the present, which does not cease to pass, and the other is the past, which does not cease to be but through which all presents pass".[19] Gilles Deleuze captures the nature of the Bergsonian contribution when he writes, "hence this consequence: Bergsonian duration is, in the final analysis, defined less by succession than by coexistence".[20] It is precisely the coexistent relations among the past, present and future that both complicate and more accurately capture temporal experience than the vector of linear time, by taking into account our layered memories and their constant presence in the now.

In his later work, *Creative Evolution*, Bergson pushes this point much further, extending his discussion of duration (and the very idea of becoming) from being anchored only in human experience to an account that carries both of these into nature itself. There Bergson suggests that duration is beyond subjective human experience and is operable in biological evolution itself, and indeed in the evolution of the universe. *Creative Evolution*, in many respects, is the true Bergsonian synthesis, the creation of a whole that is always becoming, filled with a multiplicity of vibrating intensities, marked by duration, and spurred on by a vital impulse (*élan vital*) to gnaw at the future, to create that which it stamps with its own freedom, to evolve. Here duration is extended beyond man to the whole of the living universe, which consists of coexisting degrees of contraction and expansion. The universe can then be seen as "a gigantic memory, a universal cone in which everything coexists with itself, except for differences in level".[21] When this virtuality is actualized, it is differentiated and developed; when it actualizes and develops its parts, it does so according to divergent lines. These lines are truly creative; they actualize themselves only by invention. Between plant and animal there are now only differences of degree; they all belong to the same virtual whole. The élan vital designates the actualization of these lines of differentiation; it is the vital life force that spurs creation and hence evolution. There is an element of mysticism in this virtual monism, of an open and finite God, and of souls that can discern and express this whole and find in it a virtual unity.

Bergson therefore provides us with an image of time that is true to human experience but can also be extended beyond human subjectivity to nature itself. What might be the implications of such an image of time, marked by the coexistence and interpenetration of tenses? What might be the consequences of such an understanding for our understanding of politics? As Deleuze writes, from the perspective of the Bergsonian time-image, "the question is not whether or not the past has ceased to exist but whether or not it has ceased to be useful",[22] and it is consequently to a discussion of the usefulness or the lack thereof of such an understanding that we now turn.

NIETZSCHE'S UNTIMELY PHILOSOPHY

While Henri Bergson provides us with a compelling account of the presence of the past in perception, of the coexistence of the tenses, of duration as becoming, freed from spatial representation and even operating in nature, Friedrich Nietzsche asks what this means for life itself. Bergson frees time from its restraints and exposes it as pure becoming, challenging previously held notions of sequence and linearity, Nietzsche, while diverging at times from Bergson, offers up a similarly complex account of temporality, but one that contains at its core an affirmative ethical project. As Elizabeth Grosz writes in *The Nick of Time*, the conversation between Nietzsche and Bergson is an untimely and productive one because "each provides us with the raw materials, crystals of insight, regarding the unhinging of identity and the problematization of knowledges that acknowledging the full force of temporality implies. Although they do not present a single cohesive and integrated understanding of the intrications of time and life—there are many tensions, inconsistencies, and divergences in their respective claims—nevertheless, they do represent a particular orientation that . . . enables them to interact with and affect each other".[23] There are many moments of resonance between Bergson and Nietzsche. They are both concerned with the ability to act in the present and the extent to which the past conditions such action. As Grosz writes, "For Bergson, as for Nietzsche, philosophy is an activity of incitement to action . . . It is because we act in a world whose complexity we cannot fully perceive or comprehend, which we cannot fully anticipate or control, that we invent, we make more of ourselves than life makes of us".[24]

It is in this context that history emerges as critical for Nietzsche. Conscious of the coexistence of the past with the present, he wants to explore the value of that relationship and the ways in which the historical sense can foster inertia or inspire creative action. Thus, Nietzsche writes, "[w]e want to serve history only to the extent that history serves life: for it is possible to value the study of history to such a degree that life becomes stunted and degenerate".[25] For Nietzsche, an excessive focus on a fossilized past can diminish the value of the present, and in this he shares a common understanding with Bergson of the past as inert and the present as the locus of action. He warns against a "consuming fever of history",[26] in which examples and stories of a once-great past are deified at the expense of the present, and instead urges an untimely approach. For this most untimely of philosophers what this means is "acting counter to our time and thereby acting on our time and, let us hope, for the benefit of a time to come".[27] Nietzsche's understanding of the temporal therefore emphasizes the extent to which the past is contained in the present and the effect present action has for the future.

For Nietzsche, the coexistence of the past with the present can either overwhelm and deaden or transform and create. He takes leave of structuring narratives about time and history and instead develops a tragic narrative that urges both forgetting and remembering, instinct and cultivation. Joshua Dienstag explores this tragic sensibility in Nietzsche, arguing that is "a view of human existence as fundamentally time-bound and, hence, subject to the vicissitudes of time, lacking in any permanent features".[28] It differs in kind from the characterization of time, from Socrates onwards and clearly evident in both Augustine and Kant, that order can be imposed on the fluxes and flows of time, by the pursuit of truth, the hope of redemption, or in regulative laws of reason. Dienstag argues that for Nietzsche, such a characterization of the temporal experience was "an active self-deception that made life more tolerable but less genuine. It was a retreat from a real look at time-bound existence to a pleasing fantasy of progress and happiness".[29] The tragic, for Nietzsche, lies instead in the sense that of the universe "as something constantly in flux, constantly in the process of becoming and, thus, constantly in the process of destroying . . . It recommends no cure for the pains of existence, only a public recognition of their depth and power".[30] In contrast with narratives of progress, it entails developing not laws of regulation, but an ethic of self-cultivation that requires coming to terms with the notion that the we are never completely in control of a world that is always becoming. As Nietzsche writes in the *Will to Power*, a tragic ethos is necessary to come to terms with such a world—"To impose upon becoming the character of being—that is the supreme will to power . . . Becoming as invention, willing, self-denial, overcoming of oneself, no subject but an action, a positing, creative, no "causes and effects'".[31] Coming to terms with a world of change and becoming therefore not only requires an inventive and creative will, but it also requires a complex account of time, in which the past is coextensive with the present, but in which the relationship between them is necessarily a tragic one that can account for both suffering and joy in lived experience.

"OUR TIME" AND NIHILISM: THE DEVALUATION OF ALL VALUES

Nietzsche's tragic sensibility, evident in his attendance to the mighty current of life and to the fluxes and flows of this world, makes him critical of the dominant accounts of "our time", as driven by faith in redemption or by a belief in progress. For Nietzsche such accounts undervalue life and "that which is becoming because it has no instinct for divining it . . . Thus it hinders any firm resolve to attempt something new, thus it paralyses the man of action who as one who acts, will and must offend some piety or the other".[32] Just as Bergson criticizes spatialized, linear time for imposing on duration an external sequencing, Nietzsche criticizes such accounts for the values they inspire and the resentment they foment against this world. As Walter Kaufmann writes on Nietzsche's philosophy, "Nietzsche was more deeply impressed than almost any other man before him by the manner in which belief in God and a divine teleology may diminish the value and significance of man: how *this* world and life may be completely devaluated".[33]

Consider, for instance, Nietzsche's critique of Christian time, which as we have seen in the previous chapter in the work of Augustine, views time as the progressive realization of redemption. In Nietzsche's view, "A religion which of all the hours of a man's life holds the last to be the most important, which prophesies an end to all life on earth and condemns all who live to live in the fifth act of a tragedy, may well call forth the profoundest and noblest powers, but it is inimical to new planting, bold experimentation, free aspiration".[34] Such a concept of time seems to value the moment of death, the ascent to heaven, and the much sought-after redemption infinitely more than life, which it views as a constant struggle against sin. Such a view negates inventiveness and creativity and finds its joy in the afterlife, in the realm beyond life, rather than within it. Similarly, the Enlightenment concept of time as articulated by Kant also negates the value of life. It assigns to time as a necessary, subjective postulate, a driving engine of progress and development. For modern man, this leads to an "awareness that he has to live in an historicizing, as it were a twilight mood, his fear that his youthful hopes and energy will not survive into the future. Here and there one goes further, into *cynicism*, and justifies the course of history, indeed the entire evolution of the world".[35] Such a belief in progress or in a notion of the world as driven by the motor of development diminishes the value of the individual and his creative action—it leaves little room for the exercising of plastic powers as the present is viewed simply as a mere dot along the vector of the grand force of history that sweeps relentlessly forward. Nietzsche decries the de-individualizing effects of such an account, writing that "[t]he personality and the world process! The world-process and the personality of a flea!"[36]

To Nietzsche, the twin narratives of religion and reason that structure the dominant concepts of time express and engender resentment against life. Indeed, the Christian promise of redemption and the Enlightenment faith in reason and moral imperative are purveyors of a nihilism that must be overcome. Nihilism is the belief in nothing. It is the idea that life cannot be lived joyfully if there is no external source of value. Nietzsche writes, "What does nihilism mean? *That the highest values devaluate themselves*".[37] It is the condition under which the highest values turn in on themselves by turning against life—by appealing to a transcendent God or a categorical moral imperative, by suggesting that a world without such legislative order is one of chaos and utterly deprived of meaning, where no investments of value can be made in life itself, where all enchantment has left the world. He writes, "The philosophical nihilist is convinced that all that happens is meaningless and in vain. But whence this: there ought not to be?"[38] Nietzsche is severely critical of the imperative 'ought', as it forces man to turn away from an immanent world of becoming to an adherence to a transcendent principle that legislates his existence. He writes in the *Will to Power* that "[i]t is a miserable story: man seeks a principle through which he can despise men—he invents a world so as to be able to slander and bespatter this world . . . the history of philosophy is a secret raging against the preconditions of life, against partisanship in favor of life. Philosophers have never hesitated to affirm a world provided it contradicted this world and furnished them with a pretext for speaking ill of this world".[39]

The "ought not to be" is the key source of nihilism, it draws sustenance from the history of Christianity and rationalism, and it is a view that must be overcome as it breeds passivity, resentment, and revenge. Centuries of the hegemony of monotheistic Christianity have bred the pessimism that the world itself does not have the value it ought to have, that joy exists outside of this world, and that the denial of life corresponds to the pursuit of truth. Then when doubts are posed about these sediments of cultural investment lodged in our bodily memories and language; it seems to them that meaning has been pulled out of the world altogether, when it has in fact been pulled away from confidence in one particular tradition.[40] For Nietzsche, morality is not grounded in an external command but in a gratitude for being that one cultivates for oneself. Therefore, in Nietzsche's account, life does not need to be justified by the promise of redemption; it can engender meaning through itself. In contrast, the Christian account of time is sustained by a narrative that denies life and that in the Augustinian rendition yearns to leave behind the becoming of the world for the stillness of God's eternity.

The second and related source of nihilism is reason—its attendant faith in progress and its commanding moral imperative. The two faiths of Christianity and reason are related: they are mirror images of each other; indeed, one is the outgrowth of the other. They mingle to create the nihilism of modernity and its bleakness and passivity. They breed a world of resentment in which people only see the shadows of divine purpose and command while they still retain the demand for an external source of morality. Here Nietzsche is critical of the Kantian project that necessitates a moral order in order to regulate the proper use of reason. He recognizes that Kant projects these as necessary subjective postulates rather than known conditions. But he also realizes that Kant continues to look for "signs" in the world that will speak to and redeem these postulates as such. Nietzsche writes that "[t]he presupposition that things are, at bottom, ordered so morally that human reason must be justified—is an ingenious presupposition and a piece of naiveté, the after-effect of belief in God's veracity—God understood as the creator of thing".[41] Kant's *a priori a*ssumption of a universal moral law allows him to reassert in the place of a Christian God a rational and external source by which to order and regulate life. For Nietzsche, such truths are not to be legislated universally by reason but are instead cultivated and strived for. He writes, "'Truth' is therefore not something there, that might be found or discovered—but something that must be created and that gives a name to a process, or rather to a will to overcome that has in itself no end".[42] Thus, the Kantian narrative with its teleological belief in progress gives rise to a world where transcendental principles remove man from engaging with and affirming the world in which he lives, from making full use of his plastic powers of invention and creativity.

Nihilism is grounded first and foremost in a demand to sustain an image of time that has lost much of its credibility to those who seek to retain it. Nietzsche seeks to overcome the resentment that emerges from such nihilism and to replace it with an ethos that allows one to shape the contours of one's own life and identity. The kingdom of heaven must truly form within. We must shy away from a religion of revenge and replace it with an ethics that affirms the contestability that emerges from the experience of living. It is therefore essential for Nietzsche, if he is to succeed in challenging the transcendent ideals of the Christian God or the categorical moral imperative with an immanent ethics of affirmation, to refashion the dominant image of time that these narratives sustain—to one that affirms lived experience and that savors the constant becoming and flux of life and denies eternal stillness. If nihilism is to be overcome, our understanding of time and the ethical implications of that understanding must be transformed, and it is to that end that we now turn to the most mysterious and significant element of his thought, what Heidegger called "the fundamental doctrine in Nietzsche's philosophy"[43]—The Riddle of Eternal Return.

THE RIDDLE OF ETERNAL RETURN: THE CREATIVE WILL AND THE ETHICS OF AFFIRMATION

The doctrine of eternal return forms the cornerstone of Nietzschean philosophy. It is the image of time that he formulates in order to overcome nihilism. The will must be liberated from the hangman's metaphysic that restrains it—first, from the moral imperatives and from the faith that inscribes in man an original sin that causes the will to constantly divide against itself and, second, from the idea of morality as law we must obey, even against the pressures of our lower will, our affects and passions. To overcome nihilism, to find affirmation in this world, the will must emerge as the Will to Power. The creative will must escape the linear and teleological image of time that burdens it, that denies its creativity by shackling it to a past that it cannot change, a past that it drags around like so many boulders from which only faith in a transcendental realm can provide relief. While references to eternal return are found in many of Nietzsche's works, it is in his epic *Thus Spoke Zarathustra* that the most cogent narrative is found, one that places it in the context of his critique of nihilism and the idea of the world as Will to Power.

The very nature of *Zarathustra*, which takes the form of an epic poem that is written in parable and riddle and aphorism, has allowed many Nietzsche scholars to discount the account of eternal return. Those who read Nietzsche as a romantic have often ignored it while others have construed it as a determinative principle similar to Kant's moral imperative.[44] To such critics, eternal return is only an existential ethic or test, one in which you ask what you would do if life were to return again in exactly the same form. The value of such a test, of course, is that it encourages the affirmation of life amid its tragic character, but the doctrine of eternal return is more than an existential gut check. There is much evidence that Nietzsche himself considered eternal return to be his most significant and untimely philosophical contribution, bringing together all the elements of his thought. As the Nietzsche translator and scholar Walter Kaufmann writes, "the doctrine of eternal recurrence is the most extreme repudiation of any deprecation of the moment, the finite and the individual—the antithesis of any faith in infinite progress, whether it be evolution, Faust's unbounded striving, or the endless improvement of the human soul in Kant's conception of immortality. It the antithesis too, of any faith in another world", imploring us instead to remain faithful to this earth and to not dwell in otherworldly hopes.[45] This suggests that eternal return is more than a test; it is also a fundamental philosophical doctrine about how time works.

The will to power is central to the discussion of eternal return as it is the will's suffering, its dissatisfaction with a progressive image of time, and its inability to change or transform the past that is a deep source of nihilism that in turn requires this complex and mysterious refashioning of temporal experience. As Nietzsche writes in *Zarathustra*, "*The spirit of revenge,* my friends, has so far been the subject of man's best reflection; and where there was suffering, one always wanted punishment too. For 'punishment' is what revenge calls itself; with a hypocritical lie it creates a good conscience for itself. Because there is suffering in those who will, inasmuch as they cannot will backwards. Willing itself and all life were supposed to be—a punishment".[46] It is the inability to change the past, to stem the forces that relentlessly push time forward, that inscribe in it a teleological impulse of redemption at the end of life—these are the forces that create the spirit of revenge, the bad conscience of nihilism. In these accounts, the will is a source of punishment, either divided against itself and its original sin or struggling to be in accord with a universal moral law, both of which remove from it the possibility of joy.

Nietzsche wants to refashion the creative will, to remove it from its fetters, and to cauterize its despair. Thus in a key moment he writes, "Will—that is the name of the liberator and joy-bringer; thus I taught you, my friends. But now learn this too: the will itself is still a prisoner. Willing liberates; but what is it that puts even the liberator himself in fetters? 'It was'—that is the name of the will's gnashing of teeth and most secret melancholy. Powerless against what has been, he is an angry spectator of all that is past".[47] This shackling of the will to a past it cannot change; that which imprisons it must be transformed because "this, indeed this alone, is what *revenge* is: the will's ill will against time and its 'it was.'"[48] Man must be led away from all these fables and instead be taught that "'the will is a creator'. All 'it was' is a fragment, a riddle, a dreadful accident—until the creative will says to it, 'But thus I will it; thus shall I will it'. But has the will yet spoken thus? And when will that happen? Has the will been unharnessed yet from its own folly?"[49]

This unharnessing of the will involves freeing it from its melancholy about a past it cannot change, a past that remains inert but is never actualized. It can only come from a transformation of the image of time, through the doctrine of the Eternal Return. Time is not merely a linear sequence from Creation to the Day of Judgment, nor is it the cyclical repetition of the same, as it is in the Stoic tradition. It is a knot of the past, present, and future in the moment, where what returns eternally, rather, is the rupture of the moment, in which past and future collide in the pressure of the moment. The structure of the moment returns, but what issues from the accumulation of moments is not a long cycle, but becoming itself. Eternal return represents a deeply mysterious understanding of time, as not mere cyclical repetition but of time in which the past coexists with the present and the present coexists with the future. It touches on the Bergsonian time-image here, but it goes further than the Bergsonian account we have encountered by folding into that decisive moment an affirmative will to power. Eternal return provides a powerful challenge to "our time" by presenting a complex and out-of-joint experience of time as not teleological determined by God or by a categorical moral imperative. It seeks to overcome the resentment fomented by those accounts in a moment of decisive affirmation.

In the key passage in *Zarathustra* where he first introduces the concept of Eternal Return, Nietzsche writes,

> From this gateway, Moment, a long, eternal lane leads *backward:* behind us lies an eternity. Must now whatever *can* walk have walked on this lane before? Must not whatever can happen have happened, have been done, have passed by before? And if everything has been there before . . . must not this gateway too have been there before? And are not all things knotted together so firmly that this moment draws after it all that is to come? Therefore—itself too? For whatever *can* walk—in this long lane out there too, it *must* walk once more.[50]

There are a number of elements to take note of here. First, the understanding of the moment as not merely a stage along the irreversible and progressive current of time but as knotted together with a past that it draws behind itself, that is connected to it, and that is coextensive with it and a future that it anticipates in the moment. The tenses are therefore knotted together. But precisely when Nietzsche appears to be reiterating the eternal cyclical time of the ancients, he introduces into time a profound break. Second, Nietzsche does not simply suggest that all that will happen has happened before; instead, he introduces into this account a critical note of affirmation as it is only that which *can* walk that *must* walk. As Heidegger writes, a collision between the past and the future occurs in the moment: "it occurs only to one who does not remain a spectator who *is himself* the Moment, performing actions directed toward the future, and at the same time accepting and affirming the past, by no means letting it drop".[51]

Here Nietzsche is in profound agreement with Bergson, seeing the protraction of the moment in duration as the locus of action, the moment that actualizes the past, through which it gnaws a hole into the future. However, Nietzsche adds to the Bergsonian account by introducing into the account of time and memory an ethical impulse that the past that is always present must be affirmed and transformed by the plastic power of a creative will, to use the language of his *Untimely Meditations*, to affirm time as becoming, to serve life in its flux and change, and to affirm the joyousness of that condition amid the suffering it brings. As Deleuze writes, "we must understand that Nietzsche does not recognize his idea of eternal return in his predecessors of antiquity. They did not see in the eternal return the being of becoming as such . . . They even saw it as the opposite, a subjugation of becoming".[52] Eternal return affirms becoming, not the fixed order of things, its constant flux and emergence, the compass of which is grasped in a moment that says yes. It does not belong to the realm of a stable nature governed by fixed and permanent laws; it affirms life in a world that is ever-changing and always becoming, in a decisive moment of action.

If time is eternal, but an eternity which is composed of infinite becoming, it is never the return of the same or of a constant universal such as the cyclical time of the Greeks. Deleuze explains that "[w]e misinterpret the expression 'eternal return' if we understand it as 'return of the same'. It is not being that returns but rather the returning itself that constitutes being insofar as it is affirmed of becoming and of that which passes".[53] It is not a matter of the simple repetition of the same but of a profound understanding of time as becoming, that requires a moment in which the Will affirms return, by choosing to exercise its plastic powers on the past. It is not the return of the same, but of the changed, it contains in its repetition a profound moment of difference. As Heidegger puts it, the essential thing to grasp is the action in the moment at which recurrence occurs. He writes that "that which is to come is precisely a matter for decision, since the ring is not closed in some infinity but possesses its unbroken closure in the Moment, as the center of the striving; what recurs—if it is to recur—is decided by the Moment and the force with which the Moment can cope with whatever is repelled in it by such striving".[54] The profound secret of the riddle of eternal return is that it asks us whether "we merely want to drift with the tide of things or whether we would be creators. Prior to that, it asks us whether we desire the means by which we might again *become* creators".[55]

How does thinking about time in this way, as a moment knotted together with the past and the future, a moment in which whatever can walk, must walk again—how does such an account help to overcome nihilism and the devaluing of life? Consider the dramatic turn the parable takes as soon as Zarathustra divulges the secret of eternal return. He suddenly hears a dog howl, and it triggers a memory from his childhood. He recalls this image of the dog bristling and its head trembling, just as the eternal return of time reveals itself to him. He then sees a man on the ground, and the howls of the dog became cries for help. A young shepherd, "writhing, gagging, in spasms, his face distorted, and a heavy black snake hung out of his mouth. Had I every seen so much pale nausea and dread on one face?"[56] Zarathustra tears at the snake in vain and then cries out, "'Bite! Bite its head off! Bite!' Thus it cried out of me: my dread, my hatred, my nausea, my pity, all that is good and wicked in me cried out of me with a single cry".[57] The shepherd heeds Zarathustra's word and bites violently into the snake that is choking him and emerges as "no longer shepherd, no longer human—one changed, radiant, laughing! Never yet on earth has a human being laughed as he laughed!"[58]

What does this encounter signify? Why does Zarathustra recall such a horrific sight just as the secret of eternal return is revealed to him? Who is the shepherd, and what is the snake? The vision Zarathustra sees is of man, being choked by the spirit of revenge, by the coils of the snake that is hanging out of his mouth. As Heidegger writes, "the black snake is drear monotony, ultimately the goallessness and meaninglessness of nihilism. It is nihilism itself".[59] No amount of external pulling and tugging can dislodge the snake; it must be overcome from within, in a decisive moment of action. To follow Heidegger, "the implication is that nihilism cannot be overcome from the outside. We do not overcome it by tearing away at it or shoving it aside—which is what we do when we replace the Christian God with yet another ideal such as Reason, Progress".[60] The snake's head must be bitten off; it must be destroyed from within, in a decisive moment of affirmation, and when that nihilism is conquered by a creative will, joy can be restored to the world and meaning can again be invested into its fluxes and flows.

It is because the doctrine of Eternal Return contains the key to the overcoming of nihilism that Zarathustra sees the shepherd being choked by the snake. In its internal logic of repetition with difference, Eternal Return removes the spirit of revenge by requiring a powerful ethical moment of affirmation of this life. The bad conscience cannot return because the repetition destroys it, as the creative will does not will it to return. As Deleuze writes, "the small, petty, reactive man will not return. In and through eternal return negation as a quality of the will to power transmutes itself into affirmation, it becomes an affirmation of negation itself, it becomes a power of affirming, an affirmative power".[61] To affirm Eternal Return as a nonlinear and complex account of time is to affirm life at its core; it is to overcome nihilism and the bad conscience of religion and reason. It is therefore the solution to the riddle that helps Zarathustra overcome his nausea. It does not eliminate or transcend the tragic character of life. Rather, it transmutes the tragic possibilities that inhere in a world of becoming into an affirmation of life amid both the joy and suffering of this condition. It does so precisely because it contains within it a creative will and ethos of affirmation; it requires of man not to submit docilely to accounts of life as suffering from which he must be redeemed, or life as devoid of value without a moral imperative, but to embrace the complexity of life, to see it as an infinite stream of becoming and to affirm and celebrate the multiplicity and difference implicit in that understanding. It does so even while simultaneously recognizing the suffering it engenders because of its lack of permanent stability, without the reassuring anchors of postulates and moral imperative. It requires of man that whatever happens in the moment, to will it again, but this time to will it as a creator, to transform the boulders of the past with his plastic power in the moment. As Nietzsche puts it, "As creator, guesser of riddles, and redeemer of accidents, I taught them to work on the future and to redeem with their creation all that *has been*. To redeem what is past in man and to re-create all 'it was' until the will says, 'Thus I willed it! Thus I shall will it.'"[62] The key contribution of Eternal Return is to overcome the nihilism fostered by the devaluation of values. It is to free time from the teleological narratives that structure it and restore to it the idea of time as becoming. As Zarathustra says, "'By Chance'—that is the most ancient nobility of the world, and this I restored to all things: I delivered them from their bondage under Purpose. This freedom and heavenly cheer I have placed over all things like an azure bell when I taught that over them no 'eternal will' wills".[63]

By refashioning the image of time thorough the concept of eternal return, Nietzsche provides an alternative narrative to that which sustains "our time". In contrast to the purposiveness of a heavenly creator, he posits chance, becoming, and joy. Nietzsche therefore moves away from the transcendence of Augustine and the postulates of transcendence of Kant to discover and cultivate a creative and affirmative ethics that recognizes the inescapably tragic character of a world of becoming but finds joy in its suffering. A Nietzschean ethics is one anchored in an onto-story of the moment that eternally returns, which opens up the possibility of a creative will to power that reevaluates values and finds its joy in the very complexities of life—a life that it cannot authoritatively control, a life that is tragic in character that is played out in a world of becoming. This is exemplified in the concluding fragment of the *Will to Power* where Nietzsche provides a compelling and poetic account of what such affirmation of this life and this world represents:

> Do you know what "the world" is to me? Shall I show it to you in my mirror? This world: a monster of energy, without beginning, without end; a firm, iron magnitude of force that does not grow bigger or smaller, that does not expend itself but only transforms itself . . . as force throughout, as a play of forces, and waves of forces, at the same time one and many, increasing here and at the same time decreasing there; a sea of forces flowing back and rushing together, eternally changing, eternally flooding back, with tremendous years of recurrence, with an ebb and a flood of forms . . . still affirming itself in this uniformity of its courses and its years, blessing itself as that which must return eternally, as a becoming that knows no satiety, no disgust, no weariness: this, my *Dionysian* world of the eternally self-creating, the eternally self-destroying . . . my "beyond good and evil".[64]

Nietzsche's world of becoming is revealed as a world of forces and energy that is never depleted but is always changing in larger or small ways, a world in which waves of forces play together and act and react on each other. He reveals a world of gaiety, dance, and becoming where only that which can be affirmed returns, and therefore inscribes in it a tragic ethos that embraces creativity and complexity.

This is the very essence of what Nietzsche calls "*Dionysian wisdom*"—it is the sense of "[j]oy in the destruction of the most noble and the sight of its progressive ruin: In reality joy in what is coming and lies in the future, which triumphs over existing things, however good".[65] For Nietzsche, affirmation and joy is derived from this very impermanence, this constant change in Nature, which, as Bergson would have it, is permeated by an élan vital that animates it, with varying intensities, and which invests it with duration, a world that is eternally self-creating and self-destroying. In affirming such a world, Dienstag argues, "Nietzsche is, in some sense, thereby recommending the practice of life".[66] Above all, it is the tragic sensibility that "this pattern of destruction and creation is unalterable and must be borne. And it cannot be withstood by means of faith in progress. We must learn to hope in the absence of an expectation of progress".[67] Nietzsche reveals a world that is ever changing, that is full of suffering and joy, but a world that can still by affirmed the creative will in a moment that eternally returns. In doing so, he presents a profound challenge to the ordered and regulated world of stillness that is underlined by teleological accounts of time.

Complex accounts of time, as articulated by Bergson and Nietzsche, therefore present a significant contrast to the dominant accounts of "our time". They argue that the out-of-joint experience of time in everyday life cannot be captured by the linear vector that emerges from the Augustinian and Kantian time-image. Henri Bergson, in his account of duration and memory, shows how time operates in lived experience, by eschewing transcendental accounts for the methods of intuition, in attempting to go beyond the conditions of experience. Bergson shows us that the past coexists with the present, that it colors our perceptions and conditions our actions in the moment. For him the past is ever present, but inert—it is the layered memory that must be actualized by the present which is the locus of action. In the Bergsonian image, time emerges as duration, where the tenses do not follow each other in a neat, orderly sequence but are coexistent and interpenetrated. By displacing the neat sequentiality of linear and progressive time, Bergson frees time from its spatial representations in philosophy and allows it to emerge as pure becoming, which extends beyond human experience to nature itself, which is animated by a vital impulse. Nietzsche concurs with Bergson in affirming the coexistence and nonsequential nature of time but probes the question further, asking what effect such a temporal understanding of the coexistence of the past might have on life itself and how we live it. Nietzsche examines the advantages and disadvantages of such an understanding and urges the cultivation of a tragic sensibility that requires remembrance and forgetting in equal measure. The boulders of the past can shackle us to the dictates of tradition, but if we use our plastic and creative powers, we can also act on that past in a transformative manner. In doing so, Nietzsche presents a powerful critique of the Augustinian and Kantian conceptions of time, by examining the ways in which they diminish life, by promising redemption in a world beyond this one, or by positing a categorical moral law to order our actions in this one. He argues that such concepts of time and the narratives of faith and reason that sustain them are sources of nihilism, which must be overcome. In their place he offers the complex temporal doctrine of Eternal Return that posits a world of becoming and flux, central to which is an ethic of affirmation of this life in all its complexity and the ability to act in an untimely fashion—to transform the past for the sake of the present for the benefit of a time to come, by the actions of the creative will.

If dominant accounts of "our time" are indeed eroding in our late modern world, as Sheldon Wolin has suggested, it is critical to examine what these alternative accounts can offer in their place.[68] As Bergson and Nietzsche have shown, such accounts of time splinter the neat and sequential chronology of linear time by highlighting instead the coexistence of the tenses and of time as infinite becoming. They compel us to encounter a tragic world of fluxes and flows, not regulated by a redeeming God or anchored in reason, and call on us to affirm the multiplicity and difference with which it is replete. Such complex accounts of time seek to affirmatively overcome the transcendental and determinist narratives of "our time" for a world of immanence, a world of riddles and sportive monsters. If dominant narratives of teleology and progress can no longer sustain us, in what ways might such an out-of-joint account of time, one that affirms the chaos and the contingency of lived experience, offer as an alternative? What implications might such an immanent view have on our understanding of the political?

If memories and concealed pasts are ever present, how might that inform our understandings of the power of silenced pasts, of communities haunted by memories of genocide or ethnic conflict, of pasts unacknowledged and unredeemed? To explore these questions further, it is to a broader discussion of the political implications of thinking in an out-of-joint mode of time to which this book now turns.

NOTES

1. Georgio Agamben, "Time and History" in *Infancy and History: Essays on the Destruction of Experience,* trans. L. Heron (New York: Verso Books, 1993), p. 101.
2. Gilles Deleuze, *Bergsonism*, trans. H. Tomlinson and B. Habberjam (New York: Zone Books, 1991), p. 23.
3. Henri Bergson, *Matter and Memory*, trans. N.M. Paul and W.S. Palmer (New York: Zone Books, 2002), p. 11.
4. Ibid., p. 73.
5. Ibid., p. 150.
6. Ibid., p. 153.
7. Ibid., p. 166.
8. Ibid., p. 35.
9. Ibid., pp. 133–134.
10. Ibid., p. 134.
11. Ibid., pp. 68–69.
12. Ibid., p. 137.
13. Ibid., p. 153.
14. Ibid., p. 138.
15. Ibid., p. 138.
16. Deleuze, *Bergsonism*, p. 55.
17. Bergson, *Matter and Memory*, p. 149.
18. Deleuze, *Bergsonism*, pp. 58–59.
19. Ibid., p. 59.
20. Ibid., p. 60.
21. Ibid., p. 100.
22. Bergson, *Matter and Memory*, p. 149.
23. Elizabeth Grosz, *The Nick of Time: Politics, Evolution and the Untimely* (Durham, NC, and London: Duke University Press, 2004), p. 6.
24. Grosz, *The Nick of Time: Politics*, p. 12.
25. Friedrich Nietzsche, *Untimely Meditations*, trans. R.J. Hollingdale (Cambridge: Cambridge University Press, 1997), p. 59.
26. Ibid., p. 60.
27. Ibid., p. 60.
28. Joshua Foa Dienstag, "Tragedy, Pessimism, Nietzsche", *New Literary* 35, no. 1 (Winter 2004): p. 87.
29. Ibid., p. 90.
30. Ibid., p. 87.
31. Friedrich Nietzsche, *The Will to Power*, ed. W. Kaufmann and trans. W. Kaufmann and R.J. Hollingdale (New York: Vintage Books, 1968), pp. 330–331.
32. Nietzsche, *Untimely Meditations*, p. 75.
33. Walter Kaufmann, *Nietzsche: Philosopher, Psychologist, Antichrist* (Princeton, NJ: Princeton University Press, 1974), p. 101.

34. Nietzsche, *Untimely Meditations*, pp. 101–102.
35. Ibid., p. 107.
36. Ibid., p. 107.
37. Nietzsche, *The Will to Power*, p. 9.
38. Ibid., p. 23.
39. Ibid., p. 253.
40. On this subject Gilles Deleuze has written that "[f]or Christianity the fact of suffering in life means primarily that life is not just, that it is even essentially unjust, that it pays for an essential injustice by suffering, it is blameworthy because it suffers. The result of this is that life must be justified, that is to say, redeemed of its injustice or saved". See Gilles Deleuze, *Nietzsche & Philosophy*, trans. H. Tomlinson (New York: Columbia University Press, 1983) p. 15.
41. Nietzsche, *The Will to Power*, p. 262.
42. Ibid., #552, p. 298.
43. Martin Heidegger, *Nietzsche. Vol. II: The Eternal Recurrence of the Same*, trans. D.F. Krell (San Francisco: Harpers & Row, 1979), p. 6.
44. For a detailed discussion of the treatment of Eternal Return in Nietzsche scholarship, see Kaufmann, *Nietzsche: Philosopher, Psychologist, Antichrist*, pp. 321–322.
45. Ibid., p. 321.
46. Friedrich Nietzsche, *Thus Spoke Zarathustra*, trans. W. Kaufmann (New York: Penguin Books, 1978), p. 140.
47. Ibid., p. 139.
48. Ibid., p. 140.
49. Ibid., p. 141.
50. Ibid., p. 158.
51. Heidegger, *Nietzsche*, p. 57.
52. Deleuze, *Nietzsche & Philosophy*, p. 29.
53. Ibid., p. 48.
54. Heidegger, *Nietzsche*, p. 57.
55. Ibid., p. 174.
56. Nietzsche, *Thus Spoke Zarathustra*, p. 159.
57. Ibid., p. 159.
58. Ibid., p. 160.
59. Heidegger, *Nietzsche*, p. 179.
60. Ibid., p. 179.
61. Deleuze, *Nietzsche & Philosophy*, p. 71.
62. Nietzsche, *Thus Spoke Zarathustra*, p. 198.
63. Ibid., p. 166.
64. Nietzsche, *The Will to Power*, pp. 549–550.
65. Ibid., p. 224.
66. Dienstag, *Tragedy, Pessimism, Nietzsche*, p. 91.
67. Ibid., p. 93.
68. Sheldon Wolin, *What Time is it?* in *Theory and Event* 1:1 (Project Muse, Johns Hopkins University Press, http://muse.jhu.edu/journals/tae, 1997).

4 The Politics of Memory

William Faulkner famously wrote that "[t]he past is never dead. It's not even past".[1] This continued presence of the past is evident in our daily experiences of individual memory. The memory-fragments from a previous encounter imbue our present experience with an uncanny hue and shape and shove our actions in the now. Attending to the fluidity of memory in our everyday lives entails the recognition that time is experienced as "out of joint"—where time does not seem neatly sequential, or obediently linear, where the past, present and future do not seem to follow each other in calibrated intervals. If the past is always present, and indeed if it is never fully past, how do we negotiate the politics of memory on a collective level—when the memory in question is that of a nation, or an ethnic group, or of a race? How do we reconcile the demands of an injured past for recognition and redemption with the desire of the present to leave it behind and embrace a new future?

Societies that have suffered silenced and injured pasts are faced with a series of thorny challenges in trying to address the presence of their pasts. First, they face the issue of translating thousands of individual memories into a collective memory. If memory exists in fragmentary form in individuals and can change with the demands of the present, the difficulty of coalescing such a multiplicity of memory-fragments into something of an authoritative collective memory is profound. Second, they have to determine how to address the past. These attempts can take the form of acts of public legislation, trials, amnesty, or reparations, among others. They have to negotiate a balance between remembrance and forgetting, between forgiveness and vengeance. In a collective setting, they are also faced with the impossibility of satisfying all constituencies involved. This is further compounded by the concreteness and inescapability of cultural difference, which discounts neat and comprehensive solutions to the politics of memory. Not only is it unfeasible to have a one size fits all prescription, but there is also an inherent fragmentation and multiplicity in the scale of national tragedy that can make any attempt at a collective response seem incomplete.

Negotiating the politics of memory requires navigating a difficult terrain devoid of simple answers. What is a collective memory? How might it be constructed? A collective memory can be understood as the combined memory of a population that has experienced a common past. Collective memories can take the form of authoritative accounts of the group's past, such as museums, an act of legislation that articulates an account of the past, or a "truth" commission that seeks to solidify many perspectives into a larger narrative. It may also take a more decentralized approach as in the oral history of a community that changes over time, given the participants involved, or the creation of works of art or memory cloths. The difficulty of creating a collective memory stems from the acknowledgement that it is a compilation of multiple individual memories, offered by those who have experienced a common event. Michel-Rolph Trouillot has written that "the problems of determining what belongs to the past multiply tenfold when that past is said to be collective".[2] In part, the problem of authoritatively declaring a collective past stems from the fact that individuals do not experience an event in the same way and that relationships of power often determine whose accounts are included and whose are excluded in the construction of a collective memory.

Complicating the issue further, individual memory does not function as a storage chest full of discretely preserved and intact memories that emerge whole when recalled by perception. Gilles Deleuze has written that individual memory is itself fragmentary, with shards and splinters of it embedded in layers of the past that are called up by both conscious revelation and unconscious pressures. In Deleuze's terms, there are "circles of the past constituting so many stretched or shrunk regions, strata, and sheets: each region with its own characteristics . . . depending on the nature of the recollection we are looking for, we have to jump into a particular circle".[3] Memory is not stored intact for conscious retrieval; it exists in individuals in fragments and splinters that are called up into and coexist with the present. Indeed, it is not always possible to distinguish a memory-fragment from the perception of the present. As Henri Bergson (who heavily influences Deleuze) has suggested, "if there be memory, that is, the survival of past images, these images must constantly mingle with our perception of the present and may even take its place. For if they have survived it is with a view to utility; at every moment they complete our present experience, enriching it with experience already acquired".[4]

If there is no single national memory that has been stored intact in the present, how can a collective memory be created from the memory-fragments of a multiplicity? Such a construction seems difficult given the complexity and multiplicity of coalescing thousands of individual memory-fragments. However, the challenge of articulating an authentic collective memory should not discount the political benefits of being attentive to the presence of an injured past. Despite the problems posed by questions of compilation and inclusion and allowing for different accounts of the same event, there is still significant political value in attempting to create a collective memory of a common past, particularly when the past is one of injury. For Trouillot, there is value in engaging the politics of collective memory despite the problems of accumulation and collection because for many, "their constitution as subjects goes hand in hand with the continuous creation of the past. As such they do not succeed such a past: they are its contemporaries".[5] Collective memory is essential to subject formation in the present, to how previously injured constituencies orient themselves to participation in the politics of the future. If their very identity-constitution as citizens is inextricably linked to their perception of a past that is always present and contemporaneous, the creation of and the responses to collective memory emerge as essential political questions.

In countries where the past is one of grievous injury, in which large sections of the population have been silenced by mass violence, addressing the place of past injury in the present is imperative to the politics of the future. As Michael Ignatieff has written, "These places are not living in a serial order of time but in a simultaneous one, in which the past and present are a continuous, agglutinated mass of fantasies, distortions, myths and lies".[6] In societies where the presence of the past is overwhelming, where the out of joint experience of time becomes a permanent and collective experience, it can result in a host of problems. Not addressing the mélange of pasts coexisting with the present can damage communities traumatized by violence. It can permit the perpetuation of pernicious ethnic and racial stereotypes, allow factual errors to rise to the level of myth, and let perpetrators of atrocities to walk unchecked among their victims. In short, not engaging the presence of the past can cripple a society's ability to participate in the politics of the present and the future. The politics of memory profoundly affect a society's sense of its own identity, the possibilities for civic concord, the respect for the rule of law, and its recognition of the human rights of all its citizens. Thus, while the challenge posed by coalescing thousands of individual memories into that of a collective memory is an operation that is necessarily fractured and fraught with difficulty, if countries are to achieve reconciliation or come to terms with the tragic nature of their pasts, they must attend to the place of memory in the present in order to craft a new politics of the future.

In order to examine the politics of collective memory, this chapter extends the analysis of Nietzsche's writings on time and memory developed earlier in this book, to focus on his reflections on the resentment caused by the presence of the past and on the ways in which it might be overcome. It then analyzes three recent attempts to negotiate the politics of collective memory through this Nietzschean lens—the referendum for amnesty in Algeria in the aftermath of decades of violence and civil war, the Truth and Reconciliation Commission in South Africa and its attempts to create an archive of national memory to help overcome its apartheid past, and the *gacaca* courts of Rwanda, a community-based system of justice that has emerged in recent years as one part of a tripartite strategy to heal the country from the brutalities of its genocide, one in which vast swathes of the population were complicit. Finally, it turns to an exploration of the difficulty of posing a comprehensive solution to the politics of memory, arguing instead that in a complex world of difference, such negotiations of the presence of the past, while essential for crafting a new politics of the future, must necessarily remain fragmentary and incomplete.

FORGETTING, REMEMBRANCE, AND AFFIRMATION

This chapter takes its bearings from a set of untimely reflections by Friedrich Nietzsche. Previously in this book, Nietzsche's writings were explored to unearth a complex account of time that was true to the out-of-joint experience of time that challenged the rigid sequentiality of the dominant Western account. Here Nietzsche's texts are used for a different purpose: to help formulate an analytic framework to negotiate the untimely presence of the past in collective memory. In his *Untimely Meditations*, a set of essays that anticipate many of the issues that later become central to his philosophy, Nietzsche reflects on the question of the always-present past, the tendency toward resentment it foments in the present, and the ways in which it might be overcome. But he also asks, "In what ways can an overwhelming historical sense of the presence of the past affect life?" "How can it be a disadvantage for action and affirmation in the present?" and "In what ways can it foster genuine creative impulses in the now?" In this account, Nietzsche provides helpful insights into the question of degree in negotiating the presence of the past—how much of the past to forget and how much of it to remember if it is to be truly transfigured in the present.

Nietzsche is acutely aware that man's relationship to the past is a complex one. The past is not just a trove of memories that conditions the present; it can also serve as an impediment to action in the now, paralyzing individual creativity with the sedimentation of tradition and habit. Man, in Nietzsche's view, "cannot learn to forget but clings relentlessly to the past: however far and fast he may run, this chain runs with him".[7] He contrasts man, who can be held captive by the past, to the animal who lives unhistorically—"it does not know how to dissimulate, it conceals nothing and at every instant appears wholly as what it is; it can therefore never be anything but honest. Man, on the other hand, braces himself against the great and ever greater pressure of what is past: it pushes him down or bends him sideways, it encumbers his steps as a dark, invisible burden".[8] He warned of an overabundance of history, of a consuming fever of the past that could overwhelm and encumber man and remove him from the experience of life and living and counseled forgetting as an appropriate response, "as essential to action of any kind, just as not only light but darkness too is essential for the life of everything organic".[9] To that end, the ability to sometimes forget, to ignore the demands of the past is essential for happiness. Happiness is "the ability to forget or, expressed in more scholarly fashion, the capacity to feel *unhistorically* during its duration".[10] Interestingly, recent experiments by memory researchers in Scotland seem to confirm Nietzsche's insight on the essential need to forget. Through a series of experiments they found that a "significant forgetting effect" could be achieved, which might "potentially help people with depression or maybe even post-traumatic stress disorder" by allowing them to forget the pain caused by memories of trauma or injury.[11]

When applied to collective memory, forgetting can be seen as necessary for living itself—to not dwell incessantly and bitterly in a past that one cannot will away, but to attempt to put it behind in an effort to engage with the present. But when is it necessary to forget and when is it essential to remember? Nietzsche also wrote that is essential to determine "the boundary at which the past has to be forgotten if it is not to become the gravedigger of the present".[12] It is this boundary that is always subject to negotiation and struggle in the politics of collective memory—when to forget, when to remember, and how to determine what role the past must play in the present and in shaping the politics of the future. Acknowledging the inescapable presence of the past can allow it to be made into a condition of the present and worked on in productive and transformative ways. In Nietzsche's view, such recognition could affirm life even as it recognized the tragic nature of a world without intrinsic purpose whose complexity cannot be fully comprehended or confidently controlled. The tragic relationship that he posits denies the basic opposition of the past as either destructive or inspiring. It puts forward an account that requires attention to context and to individual negotiation. For Nietzsche, the past is an "imperfect tense that can never become a perfect one".[13]

In the absence of a simple opposition, he urges inventiveness and the cultivation of the past for life, creativity, and action. The past that constantly and inertly presses into the present need not imprison it with the dead weight of history; it can be creatively called on to serve and indeed to transform the present. Nietzsche calls this creative ability the "plastic power" of man, of a people, or of a culture. He writes, "I mean by plastic power the capacity to develop out of oneself in one's own way, to transform and incorporate into oneself what is foreign, to heal wounds, to replace what has been lost, to recreate broken molds".[14] It is possible for man to draw the past into himself, to not lose himself to the past—"the most powerful and tremendous nature would be characterized by the fact that it would know no boundary at all at which the historical sense began to overwhelm it: it would draw to itself and incorporate into itself all the past, its own and that most foreign to it, and as it were transform its blood".[15] Such an ability to utilize the past could put together the pieces from the wreckage and yield strength for the future. Using the plastic power of the past could "serve life"—by transforming the injuries of the past into an affirmative condition of present action. This plastic power would require coming up with innovative and creative responses to the presence of an injured past in order to overcome it. Forgetting and remembering, the unhistorical and the historical sense, are needed in equal measure for the health of a people and a culture, but just as important is the ability to work on that past and use it to serve life.[16] Nietzsche therefore provides an important analytic framework for negotiating the untimely politics of memory. He argues that both forgetting and remembering are necessary tools in negotiating the politics of memory for the health of a society. And he argues that an injured past might be overcome by attending to the conditions of resentment such a past creates in the present and actively working on it in a manner that affirms life. In other words, Nietzsche encourages approaches and actions that seek to transfigure the weight of the past by re-crafting it in the present for the possibility of future action, even if such acts are deeply painful and prove almost unbearable.

In a significant work on cases of mass violence that is singularly attentive to the politics of collective memory, Martha Minow concurs with Nietzsche on the importance of both remembrance and forgetting and on the impossibility of finding the optimal balance between them in politics. She writes, "A common formulation posits the two dangers of wallowing in the past and forgetting it. Too much memory or not enough; too much enshrinement of victimhood or insufficient memorializing of victims and survivors; too much past or too little acknowledgment of the past's staging of the present".[17] For Minow, as for Nietzsche, this is a boundary that must always be negotiated; the perfect calculus can never be fully realized or fixed in advance by policy makers. It is worth asking why exactly it is so difficult to articulate an optimal combination of forgetting and remembering in politics. There are numerous responses to this. In a collective setting, one major stumbling block is the impossibility of satisfying all constituencies involved. If a nation decides as a whole to forget, where does that leave those who insist on remembrance and cannot find space to articulate their memories in a national narrative that is predicated on excluding them? What impact might that have on citizenship and the construction of political identity? The question is then further compounded by the concreteness and inescapability of cultural difference, where difference is often marked by power and silenced by the state. In such conditions, neat and comprehensive solutions to the politics of memory prove to be almost impossible. We have moved from the realm of Kant and Augustine, for whom a common source of morality resulted in a common image of time that made a universalist analysis of politics possible. Instead, as Sheldon Wolin has suggested, in a complex world of difference, with multiple culturally constituted experiences of time and no common concept of "our time", an attempt to come up with a universal solution to the politics of memory is something of a fool's errand.[18]

Not only is it difficult to formulate a one-size-fits-all prescription that strikes the optimal balance between forgetting the past or remembering it, but there is also an inherent fragmentation and multiplicity in the overwhelming scale of national tragedy that can make any attempt at a collective response seem incomplete. How can a single act of public legislation, which attempts to heal thousands of individual injuries ever allow for the overcoming of the past in the affirmative manner that Nietzsche suggested was necessary for transfiguration? Minow writes that such work "inevitably becomes a fractured meditation on the incompleteness and inescapable inadequacy of each possible response to collective atrocities".[19] This is because "the variety of circumstances and contexts for each nation, and indeed each person, must inflect and inform purposes in dealing with the past".[20] It seems rather specious to posit a universal policy solution to the question of the degree to which the past can be engaged; to somehow enshrine in the human rights regime the perfect calculus of remembrance and forgetting that can be applied to each case of a ravaged and war-torn past. It is perhaps even more difficult to offer a prescription for transfiguration, such that thousands of individuals can overcome their pasts through creative acts of affirmation in the present. The manner in which these pasts are experienced, perceived, and recalled in memory by those who must negotiate them point to the implausibility of such a comprehensive solution.

Complicating the question further, the cultural prism through which methods such as criminal trials, amnesty proposals, and reconciliation hearings are viewed point toward the need to respond to the politics of memory in creative and multifaceted ways. In the absence of an overarching solution, it is necessary to focus on different, and partial, approaches. In doing so, it is important to be creative in their formulation in order to knit together responses that address each concrete situation and tries to somehow come to terms with its almost bewildering complexity. Critical to the politics of memory is to find a balance between vengeance and the desire for punitive action, on one hand, and forgiveness or amnesty, on the other hand, which often amounts to a denial of the presence of the past by an entire collective body. Such a balance requires articulations of plastic power—the ability to transform this wounded past by the creative will in order to change the future. It requires transforming the impulse towards revenge in a moment of creative affirmation, of resisting absolute forgiveness but attempting a form of reconciliation that serves life in the present. To use Nietzsche's language, it requires transfiguring the gnashing of the will against an injured past which cannot be changed into something much more affirmative. Or at least if that deeply existential threshold is deemed too unbearably high for the messiness and bluntness of legislative politics, it requires an effort to create the conditions under which such acts of individual, and existential, transfigurations might become possible.

ALGERIA'S AMNESIA: THE DESIRE TO FORGET

If the politics of collective memory involves negotiating the boundary between an abundance or a paucity of the past in an attempt to overcome it, Algeria chose the latter. In an attempt to put behind the traumatic and brutal legacy of its decades-long civil war, Algeria officially chose to forget its collective past. In Nietzsche's terms, it realized the disadvantages of history: it felt its past dragging behind it like so many heavy boulders, and refused to allow the legacy of the past to be the gravedigger of its present. This collective act of forgetting took the form of the National Charter on Peace and Reconciliation, passed overwhelmingly in a referendum on September 29, 2005. While it is important to note that the charter was widely criticized by human rights activists and families of the thousands who disappeared during the conflict, it nevertheless represented one approach of negotiating the presence of the past. The National Charter, which was championed by Algeria's President Bouteflika, was rooted in a desire to "move on", to embrace the future, and to let go of the past. In trying to figure out how to move beyond a brutal past, Algerian officials were "essentially asking their people to forget".[21]

Algeria has emerged from a brutal civil war that left more than one hundred thousand dead and thousands more still "missing". Its civil war began in 1991 after the country's first multiparty elections for Parliament, when the army stopped the voting in the first round after the fundamentalist Islamic Salvation Front won a commanding lead. In the ensuing bloodbath, militant Islamic groups attacked government agents and slaughtered civilians, while state security forces abducted thousands of people and state-armed militias dispensed vigilante justice.[22] The National Charter for Peace and Reconciliation sought to let go of this violent past in a number of ways. First, the legislation proposed to pardon Islamist militants in prison, those on the run, and even current fighters, if they agreed to lay down their arms. They were then to be indemnified from trial, even from acts of murder, unless they were implicated in massacres, rapes, and bombings in public places. Second, it shielded the army and the security forces from criticism and accusations by the families of the thousands of Algerians who disappeared during the conflict. Third, it offered financial compensation to families who have faced such outcomes, though it was made clear that it would not seek to discover the truth about their "missing" family members.[23] In fact, it explicitly stated that "the sovereign Algerian people reject any allegation aimed at holding the state responsible for the phenomena of the disappeared".[24] Official figures suggested that 80% of voters backed the Charter in the national referendum, but human rights groups and local journalists strongly questioned the veracity of those numbers and argued that they "bore little resemblance to reality".[25]

Algeria's national charter therefore sought to achieve reconciliation and to put to rest the weight of the past on the present with a combination of amnesty, indemnity, and denial (particularly in the case of the role played by the army during the conflict). While some Algerians say they would have preferred something closer to South Africa's Truth and Reconciliation Commission, to many "in this exhausted country, eager to turn the page, the promise of peace and reconciliation" was "hard to turn down".[26] Algeria's negotiation of the politics of memory did not focus on healing the wounds of an injured past. It pursued the promise of a better future by lifting the weight of the past from the present. Such an act of forgetting was seen as a patriotic act, a symbolic gesture that allowed Algeria to pursue a better future once unencumbered by its past. An official release following the referendum of the Charter the Algerian Embassy in the U.S. put it this way—"pursuing 'national reconciliation' means that people will have to put their patriotic duty first and contribute to ending the national tragedy. The charter recognizes the difficulty grieving families of individuals still unaccounted for may have in accepting the referendum. However, it explains that this is the best way to overcome the past and construct a peaceful society for the coming generations".[27] Above all, attempting to put the past to rest through the collective act of a referendum signaled to many that Algeria could now "move forward and prosper with liberty and dignity".[28]

Some Algerians argued that letting go of the past can urge society to confront issues it must resolve, like the need to improve its education system.[29] It is clear, however, that what the charter did not offer was either "answers or accountability"[30]—it did not establish an authoritative account of the collective memory, it did not call for truth-telling, and it did not set up prosecutorial trials through which the victims confront their oppressors. This has prompted many human rights organizations, opposition groups and victims' family members to criticize it as "at best, a half step toward reconciliation".[31] While the Algerian government is optimistic and convinced that it has finally put its past to rest, the years of killing have left many Algerians "alienated from one another, distrustful of their own government and locked in their own accounting of the past. Without any formal process of truth and reconciliation, the details of Algeria's history depend on who is talking".[32] There is a real concern that in the absence of reconciliation, wounds will remain unhealed and the politics of vengeance will continue, as the injuries of the past continue to make their presence felt and seek redress.[33] As recent scholarship has shown, it seems that rather than achieve reconciliation and progress, the charter has instead led to increased repression by the state of those who criticize it. It also "ran counter to President Abdelaziz Bouteflika's promise in 2005 that reconciliation would not mean repression".[34] Indeed, it could be argued that, "in some respects, the charter legislation may even have further entrenched violence inside Algerian public and political life".[35]

For some Algerians, the charter provided an opportunity to free the present from the clutches of the grave-digging past, but to others it was simply an attempt to forget a past that remained irrevocably present, which had not been reconciled and whose wounds have not healed. Thus, on one hand, Algeria offered one response to the politics of memory—to leave a calamitous past behind through conscious and communal act of forgetting. On the other hand, it seems to have done little to create the conditions of possibility that might lay the groundwork for the overcoming of the past. It did not take steps to address the "spirit of revenge" that the past exerts in the present—a vengeful impulse which has been viscerally evident in the cycles of internecine violence that have wracked Algeria. It seems unlikely therefore that collective forgetting through an act of public legislation is sufficient to come to terms with the injured past for several reasons. First, the legislation urged many of the families of the disappeared to turn the page on that chapter of their past while barely acknowledging the difficulty of their loss and compounded it by indemnifying those responsible. Second, by offering large-scale amnesty without a framework of accountability, it tried to wipe away the wounds of the past without attending to the conditions of resentment they engendered in the present. Third, with the comprehensive and final nature of the legislative act, it effectively eliminated any public space for those who might seek to work on their fractured memories in contrary ways, through political participation or activism, or individual acts of protest. Essentially, Algeria hoped that turning the page through a symbolic act of legislation would be enough to put to rest the legacy of the past. In doing so, it acknowledged the inescapable presence of the past, but in prescribing an officially sanctioned amnesia as a response, it paid short shrift to the multiplicity and complexity of collective memory and certainly did not provide conditions under which its citizens could work on overcoming the injurious past through creative acts in the present.

SOUTH AFRICA'S TRUTH AND RECONCILIATION COMMISSION: THE REDEMPTIVE PAST AND COLLECTIVE MEMORY

If Algeria's National Charter could be characterized as flawed because of its insistence on turning the page and leaving the past behind, South Africa's Truth and Reconciliation Commission (TRC) was lauded by many in the human rights movement and by scholars of international law as one of the best and most comprehensive collective responses to the presence of an injured past. The TRC, which was officially created in 1995, attempted to come to terms with the country's legacy of apartheid in a number of ways. First, it emphasized truth telling of past horrors in order to foster reconciliation in the present; second, it offered immunity from prosecution through the explicit asking of forgiveness from the victims by the perpetrators; third, it offered modest reparations; and fourth, it sought to create an archive of collective memory as an official national record of the historical truth. Thus, the main objective of the TRC was "to deal with the thorny issues of 'historical truth', on the one hand, and amnesty, reconciliation and reparation on the other".[36]

While the work of the TRC has been widely praised as offering a model example of coming to grips with the legacy of a difficult past, it has also been criticized for exacerbating racial tensions in the country. Some "vehemently reject the conjecture that 'truth' can somehow lead to reconciliation, claiming instead that uncovering the details about the horrific events of the past only embitters people, making them far less likely to be willing to coexist".[37] In contrast to Algeria's National Charter, which suggested that only by forgetting the past could the future achieve its potential, South Africa's Promotion of National Unity and Reconciliation Act, which created the TRC, made explicit the notion that addressing the injuries of the past and establishing an authoritative historical truth for collective memory could have a salutary effect and "prevent a repetition of such acts in the future".[38] It sought to "promote national unity and reconciliation in a spirit of understanding", which would transcend "the conflicts and divisions of the past".[39] It also hoped that the very act of recounting past injuries in the form of truth telling could restore "the human and civil dignity of such victims by granting them an opportunity to relate their own accounts of the violations of which they are victims".[40] In other words, South Africa chose to remember its past through the creation of an authoritative national memory and through acts of truth telling. It also hoped that the truth and reconciliation hearings themselves would actively work on the resentment of the past and help overcome it through the healing power of confession.

Central to the TRC's objectives were several key assumptions. The first assumption was that acknowledging a painful past is inherently redemptive and that it can be healed in the present through acts of truth telling. The second assumption was that the production of an authoritative account of national collective memory could be accomplished without silencing and excluding historically marginalized constituencies. The third assumption was about the feasibility of constructing a national archive of collective memory, given the difficulty of forging one authoritative collective memory from the memory-fragments of thousands of individuals.

Reconciliation typically means acceptance, apology, and forgiveness. But in South Africa, led by Archbishop Desmond Tutu, reconciliation also had significant religious overtones of healing and transformation through the charitable act of forgiveness.[41] Can truth telling contain redemptive possibilities? Can it help overcome the spirit of revenge created by the injured past? The TRC certainly hoped that such official acts of remembrance would have a therapeutic quality that could heal the rifts in the present. Martha Minow makes this point when she writes that

> with the aim of producing a fair and thorough account of the atrocities, a truth commission proceeds on the assumption that it helps individuals to tell their stories and have them acknowledged officially. Also assumed here is the premise that a final report can create a framework for the nation to deal with its past. Echoing the assumptions of psychotherapy, religious confession, and journalistic muckraking, truth commissions presume that telling and hearing a truth is healing.[42]

There is an assumption in such official acts of remembrance that individual acts of remembrance and transfiguration will translate into a national moment of healing and redemption—an assumption which might not hold up to scrutiny. What might, however, be created by such official and formalized acts of truth telling is the potential for individual moments of overcoming, which, when exposed to the wider population might have a salutary effect. As Minow writes, "the arduous process of working through trauma that individual therapists can create is not created by a truth commission, but the commission can offer therapeutic moments".[43]

South Africa's TRC attempted to negotiate the collective memory of its past in two ways. First, it chose to remember its past, and it chose to do so through the official establishing of a collective memory that would articulate the historical "truth" of apartheid. Second, it chose to create the conditions for overcoming the resentment of the past through acts of truth telling. Both aspects of the truth and reconciliation process merit closer attention. What is a collective memory, and how can it be established and accepted as true? It can be defined as "an accepted version of the truth about the country's past. By establishing a collective memory, it becomes difficult (although not impossible) to deny that certain activities took place".[44] Establishing an authoritative account of collective memory has the potential to put to rest doubts, rumors and myths by consolidating an account of historical truth. But there are a number of difficulties with the very enterprise of attempting to establish a collective memory. First, there is the possibility of systematic bias engendered by who decides to participate in truth telling—"who is willing to testify, who thinks his or her suffering is worthy, who is willing to come forward as a victim, and who is willing to accept responsibility as a perpetrator".[45] Second, there is the very feasibility of establishing a collective memory, because of the layered and fragmentary nature of memory itself. But as we know from Henri Bergson, memory is not a chamber where shards of the past are preserved intact, waiting for conscious retrieval. Rather, memory is contracted into the present in response to the pressures of the moment and persists in the present to help shape its actions. This is not to say that individual memory cannot play a role in the production of a historical truth—indeed, such memory serves an indispensable role in the subject construction of the citizens who could make up a democratic South Africa. However, the proposition of constructing an authoritative and collective "truth" from individual memory is a dicey one. If individual memories themselves are fragmentary and emerge in response to acts of recollection in the present, how can they be collected and compiled into an intact memory trove that purports to accurately represent the entire memory of a nation? Which fragments recorded in official hearings and courtroom settings will be included, and what will be ignored? The very act of compiling a collective memory is inevitably shot through with operations of power. As Trouillot has written, the production of "any historical narrative is a particular bundle of silences . . . that silencing is also due to uneven power in the production of sources, archives and narratives".[46] In this instance, it would not be remiss to recall Foucault's famous words, that "'truth' is linked in a circular relation with systems of power which produce and sustain it, and to effects of power which it induces and which extends it".[47]

How then has the TRC process, which promised reconciliation through working on a past that it assumed was necessarily redemptive, actually reinscribed certain forces of power that operated during apartheid? In some cases, those who were silenced before continued to be marginalized from the production of the collective memory and those who shielded themselves from being tarred with the brush of complicity due to their status in society also continued to avoid having their culpability written into the historical record. This is clearly seen in the case of the elite members of the apartheid regime in South Africa. As Minow notes, "the TRC also invited members of the business community, journalists, and judges to offer submissions for amnesty for their complicity with apartheid, but received almost no responses".[48] As part of a process that required voluntary acts of truth telling to redeem the past, the intentional silence of the elite meant that their abuses of power were not recorded into this official account of South Africa's collective memory of apartheid. One could choose not to participate in the truth-telling process, and many whites in positions of power simply availed themselves of that option. In the case of the TRC, the power relations that governed South African society were reproduced through the historical narrative that allowed collective memory to be produced, regulated, distributed, and circulated in particular ways.

This is evident when the TRC's exclusions are examined and are particularly glaring in the case of women whose experiences under apartheid are underrepresented in the official record. In an interesting piece on the silences of women in the TRC process, Cheryl McEwan looks closely at the TRC experience for black women who have been historically marginalized by colonialism and apartheid and now found themselves excluded from official accounts of South Africa's past. McEwan argues that the silencing of black women in the production of historical truth is a consequence of networks of power in South African society. In particular, she notes that "legal rights and public participation seem to guarantee women's political and civil citizenship in post-apartheid South Africa, but official national mythologies are deeply fractured by social practices and discourses. Pervasive sexism constantly undercuts women's equal rights and progressive legislation often seems far removed from the lived experiences of many women".[49] These networks of discursive power had the effect of silencing women so that when black women did testify in the TRC process, they spoke largely about "abuses to male relatives (sons, brothers, husbands, and fathers) . . . In addition, the media and the Commission portrayed women as 'secondary victims.'"[50] There were also "profound cultural . . . pressures placed on women that deterred them from testifying. These include discourses of shame, collusion and complicity that prevent women from discussing in public the sexual violence that is privately acknowledged to have been widespread",[51] all of which raised questions about the "gendered truthfulness of the apartheid history told through the TRC report".[52]

The TRC's attempt at creating a collective memory archive was an exercise in the production of truth which, for the reasons outlined earlier, remained necessarily incomplete. It excludes the accounts of women who have been consistently marginalized through the country's history. Numerous accounts of sexual assault and rape are underrepresented in South Africa's official rendering of collective memory and have to be sought elsewhere—such as in the recent creation of "memory cloths" that allow women to portray these memory-fragments that are not part of the official record. This gendered exclusion was also compounded by the surprising degree of protection for apartheid leaders who simply chose not to participate in the voluntary proceedings of the TRC and whose offences could therefore not be entered into the record. The official account of collective memory therefore reinscribed and perpetuated historical silences because those in power could shield themselves from being written into the record and those marginalized constituencies with violent tales of trauma could not adequately participate in official forums of truth telling. It raised fundamental questions about how collective such an official account of historical memory could be and about how representative it was of the apartheid past.

However, despite concerns about the various shortcomings of such a project, the key question remains: Might the construction of a national archive of collective memory help South Africa overcome its injured past and lead to the kind of reconciliation in the present, which can create the conditions of an affirmative politics to emerge? While the TRC has been widely praised for exploring and accounting the past in depth, it is difficult to gauge whether it has led to reconciliation. There have been numerous act of public forgiveness and a noted absence of vigilante violence. But has the apartheid past been overcome or transfigured into an affirmative present? South African officials hoped that official acts of truth telling and that establishing a national memory would allow the past to play a redemptive role in the present. The national efforts to remember and recall the past to achieve reconciliation are unparalleled in their scope and breadth, but there remain concerns about the sufficiency of the means by which they chose to negotiate the politics of memory. While they pushed for an official and widespread project of remembrance, they also insisted on public forgiveness, which had unintended consequences. Perpetrators who voluntarily stepped forward to tell the truth about past crimes were officially forgiven, sometimes regardless of whether their victims chose to forgive them and, more egregiously, sometimes before their victims were even aware that such hearings were taking place.

South Africa's TRC clearly illustrates some of the difficulties and possibilities raised by the politics of memory. In negotiating the Nietzschean boundary between forgetting and remembrance, South Africa, in contrast to Algeria, officially and collectively chose to remember its past and to forgive those who confessed to acts of injury. It also hoped that the process of remembrance itself through the multiplication of individual acts of truth telling would have a therapeutic and transformative effect for the country as a whole. By most indications, the TRC has been successful in defusing tension and the outbreak of violence and establishing some degree of reconciliation between the parties. However, in its attempts to produce an authoritative historical truth, it has risked reproducing many silences of the past, and in its desire for unity, it has paid insufficient attention to the fragmentary nature of collective memory. In pushing for widespread forgiveness and the construction of an official account of its past, South Africa has indeed made significant steps towards addressing the thorny issues of collective memory. However, by choosing to deal with the complex and fragmentary politics of memory through a universalist approach that sacrificed dissonance and density for unity and coherence, it may have omitted a necessary dimension to overcoming its past, as it reinforced the silencing of some marginalized constituencies even as it fostered reconciliation among others.

RWANDA'S *GACACA*: MEMORY, JUSTICE, AND COMMUNITY

If South Africa's TRC encountered problems because of its universalist approach, which demanded the production of an authentic collective memory and risked reproducing injured identity and historical silences, Rwanda's tripartite attempt to deal with the legacy of its past might provide instructive clues about coming to terms with the nature of memory by posing responses that are as fragmented and interwoven as memory itself. In the aftermath of its horrific genocide, Rwanda took a multifaceted approach to negotiating the politics of collective memory—combining an international war crimes tribunal, which operated with all the safeguards of international law, along with trials conducted through the national system of justice in Rwanda, and finally the (re)introduction of *gacaca*—a traditional system of courts that amount to little more than community meetings in towns and villages where perpetrators of horrific crimes come face-to-face with their victims in front of the community they seek to rejoin.

The Rwandan genocide was the most brutal and efficient killing spree in history; in its aftermath, there remained few people who did not have some degree of complicity in the mass slaughter. The sheer scope and scale of the genocide in Rwanda means that there is hardly a family remaining in Rwanda today who did not either lose someone to the violence or include someone who participated in it. As Samantha Power writes in *A Problem from Hell*, "The Rwandan genocide would prove to be the fastest, most efficient killing spree of the twentieth century. In 100 days, some 800,000 Tutsis and moderate Hutu were murdered".[53] Technology and crude weaponry combined to foment a disastrous conclusion in Rwanda, as the local radio station, Radio Milles Collines, broadcast names, addresses, and license plate numbers of Tutsi and moderate Hutus to death squads armed with crude weapons—knives, machetes, screwdrivers, and clubs with nails. The pervasive nature of the genocidal campaign, which affected all segments of Rwandan society, meant that the past remained irrevocably and unbearably present in Rwanda. Nearly everyone lost a family member or a friend to the genocide, and massive numbers of people were involved in carrying out the murders had to be prosecuted, punished, or reintegrated into society. In Rwanda, genocide survivors include "legions of children rendered orphans by the violence, children born of rape, families displaced from their homes, and witnesses to unspeakable atrocities".[54] This left the country with the "task of figuring out how perpetrators and victims could live side-by-side while the smell of death hung in the air".[55]

The scale and the scope of the genocide in Rwanda resulted in the need to address the overwhelming presence of this past in a variety of different ways. Each solution was a partial one, but together they compose a layered and significant response to the politics of memory. The first measure undertaken in Rwanda was the creation of the International Criminal Tribunal for Rwanda (ICTR), which represented the international community's belated response to hold the most serious perpetrators and high-ranking officials in government responsible for their crimes. A key aim for the tribunal was to involve the international community in the process of accountability and therefore avoid the perception of "victor's justice". It was hoped that its UN mandate would allow it to gain access to fugitives in foreign countries and would publicize the idea that the genocide in Rwanda constituted a crime against humanity as a whole.[56] While the ICTR successfully prosecuted a number of significant cases resulting in a production of a growing body of international case law on genocide and crimes against humanity, it was plagued by a number of problems and encountered considerable criticism from Rwandan society. First, its location in Arusha, Tanzania, put it outside Rwanda's borders and at a distance from Rwanda's population, which was often not aware that trials were even taking place. Second, its refusal to impose the death penalty seemed unfair to many in light of the fact that those tried in Rwanda's justice system received the death penalty for lesser crimes. Finally, the slow pace of the trials themselves, which seemed to offer criminals "deluxe justice", all led to a growing sense that the ICTR by itself was not a sufficient response to the aftermath of the genocide.[57]

The second approach involved prosecuting the accused in Rwanda's judicial system, which bore the "brunt of the genocide trials".[58] However, the judicial system itself was in a state of disrepair, as nearly two-thirds of Rwanda's judges and attorneys had either fled or been killed during the genocide.[59] In Rwanda, where so many were complicit on so many levels, there were too many suspects to deal with, and the overcrowded penal system and a dilapidated judiciary were severely taxed by this load. The Rwandan Patriotic Front rounded up close to 120,000 Hutu suspects and imprisoned them in facilities designed to hold about 40,000—"most suspects, never formally charged with any crime, were forced to live in hellish conditions: underfed, drinking dirty water, crammed into tiny rooms where they slept on top of one another in latticework formations".[60] Dealing with such a backlog would have taken the local court system an estimated two hundred years. In the meantime, the International Tribunal in nearby Arusha, Tanzania was only prosecuting high-level offenders and at times seemed more interested in establishing legal precedents for the newly established international criminal court than in the efficient prosecution of perpetrators.

It was this confluence of factors that led to the creation of *gacaca*, "the innovative third arm of the tripartite legal response".[61] The Rwandan government decided to release many low-level suspects who had confessed to their crimes to be tried in communal courts known as *gacaca*—derived from the Kinwarda word meaning "on the grass"—which referred to the setting in which these hearings took place, many of which consisted of "little more than villagers clustered around a tree".[62] *Gacaca* is founded on the principle that "the community should re-integrate the individuals whom it punishes".[63] Traditionally, *gacaca* was used to settle land disputes, but in present-day Rwanda, it was hoped that it would meet the needs of speed and inclusiveness. The trials began in 2001 and finally concluded on June 18, 2012.[64] In contrast to the trials conducted by the ICTR and the national judiciary, the *gacaca* trials took "place in and involve the communities where the atrocities were committed"[65] and were held "in front of those most directly affected—the victims, the families of victims, and the communities for whom reconciliation is not an abstract goal but a constant daily struggle".[66] *Gacaca* is rooted specifically in Rwanda's cultural tradition and requires that the very communities who have been grievously injured in its genocidal past labor to come to terms with the perpetrators themselves. They then decide whether to reintegrate them into their midst. *Gacaca* appeared to participate in a more tragic understanding of the past in which there were no assumptions that the past is either necessarily redemptive or that it is always harmful. Instead, there was a recognition in the *gacaca* process of the need to "serve life", as Nietzsche put it—to struggle with an injured past and to act on it in the present in an affirmative attempt to overcome it. *Gacaca* might well have served as an exercise of the plastic power of the community—its attempt to rework the old, to make sense of the past in the present, and to work on it for the sake of the future.

Those convicted in *gacaca* proceedings often receive reduced sentences. Some argued that was justice on the cheap—a swift way to deal with the backlog of cases by a severely traumatized population, dredging up difficult details that are too much to bear. It is argued that "handing the key processes of justice and reconciliation to a wounded population is a huge gamble".[67] Others, however, argued that there were "distinct virtues to involving the population in the trial process. For the first time, Rwandans are drawn together to discuss their problems, to publicly engage and debate with one another, and to build for the future".[68] While *gacaca* came under severe criticism from many in the human rights community because of lax legal standards, it still represented an innovative approach to the politics of collective memory. *Gacaca* trials looked very different from the trials conducted by the ICTR and the national judicial system. Defendants had no access to a lawyer and faced judges who doubled as prosecutors. Significantly, the trials took place in communities where the crimes allegedly took place and in front of a crowd that normally included the victim or the victim's family and friends. After the charges were read, the defendant was allowed to respond with his or her version of events and to call witnesses from the assembled community. If no one in the community had any evidence against the defendant, he or she was freed. The defendant could also enter a guilty plea to reduce his or her sentence.[69] In the *gacaca* model, participation in the proceedings was "seen not just as a means to an end, but as an end in itself".[70] Participation by members of the community who must decide whether or not to reintegrate the accused was seen as empowering act and the "route to building peace".[71] It was hoped that participation will "have a cathartic element for the victim or victim's family whose suffering will be publicly acknowledged . . . [that it will] lead to greater social acceptance of the trials and facilitate the integration of those found innocent . . . [and that it will allow] Rwandans to experience a different sort of mobilization that is geared at rebuilding, not destruction".[72]

Given this multifaceted set of goals, and the loose and informal structure of the *gacaca* trials themselves, it was not surprising that these meetings on the grass often took on different forms and offered differing meanings for different people. Was this process truer to the fragmentary and complex nature of memory itself? Sometimes the meetings are fairly orderly, as in the case of a meeting by a lake in 2005 where a crowd of people—a quarter of whom were amputees whose limbs had been hacked off during the genocide—sat in judgment of a suspect who confessed to having killed his neighbor's wife. In that instance, a confession was recorded, and the suspect begged for forgiveness and was asked questions by the assembled community about his crimes.[73] Other hearings followed—"one by one, people stood to tell what they knew. Sometimes there was disagreement: No, that man couldn't have killed on that day; my brother saw him in another village".[74] The community therefore came together to confront the twists and turns of its past, to mesh together memories in a public setting until a likely account was established, and then to sit in judgment of the accused.

Other hearings were less orderly and much more traumatic and raw. Here painful details were dredged up that often proved unbearable, such as in the case of one defendant who refused to allow a woman to give evidence against him and shouted through the entire hearing and leveled accusations at those present, saying that "everyone here today should be on that list of killers".[75] On other occasions, *gacaca* provided an opportunity for individuals to grapple with grief, to memorialize family members and friends. In one instance, several women brought photographs of loved ones who had died in the genocide. A journalistic account of the proceedings describes it as such:

> They clutched these photographs tightly throughout the hearing and pointed to them when they stood and gave evidence. When these women sat down again, many of them cried and hugged each other. Older women moved from the fringes of the gathering to comfort them. The women holding the photographs appeared to gain solace and strength from those who showed their concern, sitting up more confidently and soon participating again in deliberations . . . For these women, *gacaca* provided a memorial to their loved ones and perhaps some sense of healing.[76]

In such an instance, *gacaca* memorialization allowed for the expression of individual grief, and also allowed other members of the community to help those who were suffering.

In the first account described earlier, *gacaca* allowed the community to stitch together its fractured memories into a collective whole in order to decide whether to readmit a confessed killer into their midst. In the second account, it forced the community to confront a past that still unsettled and disrupted the present, when a suspect refused to hear evidence against him and accused the entire assembly of having blood on its hands. In the third account, it allowed for public memorializing that involved the community acknowledging the grief in its midst. In all these ways, *gacaca* compelled local communities to come to terms with an injured past, through acts of public memorializing, remembrance, participation, and accusation. In such hearings the past remained tragic—it was never simply redemptive or necessarily disruptive but was a complex combination of both, which must be accounted for in the present.

In contrast to South Africa's TRC proceedings and its attempts to establish a national collective memory, *gacaca* was decidedly less universalist in its approach. The community came together to address the wounds of its past, recognized its depth and power, and hoped that such proceedings might lead to some degree of reconciliation. *Gacaca* therefore was "concerned with much more than legalities. It gave people a chance to talk about their emotional experiences and for the community to acknowledge their pain and suffering".[77] However, unlike the TRC, widely praised by the international community, "*gacaca* has met with a great deal of skepticism internationally, but it is seen domestically as a hope and a possible salvation".[78] While it lacked the particulars of due process and the formalized settings of the TRC hearings, it offered something that the universalizing project of the TRC could not.

By attempting to negotiate the politics of memory in a manner that itself approached the fragmentary, raw, and disruptive nature of memory, the *gacaca* may come closer to approximating the tragic dimension of lived experience. That is, it may come to approximate how the tragic conflicts and excesses of the past must be transfigured if the parties are to have a good chance to avoid a repetition of such acts in the future. Also, in direct contrast to the TRC, the *gacaca* trials did not assume that the very act of remembrance of the past is necessarily and uniformly redemptive or that forgiveness is the ultimate goal of such proceedings. Instead, the *gacaca* process seemed to entail a recognition that such an injurious past must be worked on, that it must be reconfigured through both individual and communal acts, which cannot be proscribed in advance by the state. *Gacaca* certainly does not represent a perfect solution to be replicated elsewhere, though the model is being tried in various forms in Uganda, the Democratic Republic of Congo, Sudan, and Timor-Leste.[79] Indeed, taken as a whole, the results of the *gacaca* process appear to be decidedly mixed. Two surveys taken by Rwandan officials suggest a mixture of hopefulness and skepticism about the process. A survey undertaken in 2003 underlined this fracture clearly. In the National Unity and Reconciliation Commission (NURC) survey, 71% survivors identified "regaining peace of mind" as a key motivation for their participation in hearings.[80] However, 91% of survivors believed "that gacaca will intensify levels of trauma".[81] An additional 55% claimed they had "already suffered too much to want to participate".[82] A subsequent NURC survey taken in 2007 found that 97% of the general population and 83% of prisoners believed that *gacaca* "will eradicate 'the culture of impunity'".[83] However, it also found that almost two-thirds of the population believed that witness accounts on both sides could not be trusted and that overall trust in the *Inyangamugayo*, the community elders who conduct the hearings, had eroded significantly among both prisoners and survivors in recent years.[84] As Philip Clark has noted in his extensive study of the *gacaca* trials, "Some communities are reaping the benefits of gacaca, following constructive engagement between parties during hearings that have allowed all social groups to discuss their experiences of the genocide and fruitful interactions beyond gacaca. In such places, profound results including healing, forgiveness and reconciliation are certainly possible. Elsewhere, however gacaca has had either minimal impact, because of piecemeal community participation during hearings, or has aggravated people's circumstances through a lack of justice, inadequate degrees of truth and increased trauma".[85]

Despite its inadequacies, in a country with a long history of community-based justice and a rich oral tradition, *gacaca* offered some hope for reconciliation. By accumulating fragmentary responses to fragmentary memories and by attempting to negotiate it in different ways, it avoided some of the problems of the TRC. It offered a grassroots response that allowed for individual moments of affirmation and overcoming to occur within a collective context. Indeed, *gacaca* folded more contingency and greater complexity into its operations, by allowing individual citizens to craft their own responses within a relatively loose judicial framework in a communal setting, which led to a variety of different responses to the past, as the three accounts of the trials described earlier demonstrate. In doing so, it perhaps came closer to Nietzsche's prescription for transfiguring the past than the other solutions described in this chapter. *Gacaca* may therefore suggest something beyond either forgetting or remembrance. In deciding whether to reintegrate perpetrators of heinous crimes back into their local communities, it sought to transfigure the past, to allow it to play a positive role in the movement of the present into the future. As Nietzsche put it, our relationship with the past is always a tragic one, which can be understood as one in which the complex workings of time are not fixed by providential design in a manner that can understood and mastered by us, but one where the past impacts the present in different ways that are neither completely redemptive or paralyzing. Such a tragic understanding of the past requires coming to terms with the fact that the worst sometimes happens to us and marshaling enough courage and love of living to be able to go forward into the future without resentment. The complex and creative response of *gacaca* to Rwanda's horrific memory of genocide allowed both individuals and the community to come together in the present to address the injured past in multiple and fragmentary ways for the sake of the future. *Gacaca* provides hope that the transfiguration of the past, however partial, is possible.

NEGOTIATING COLLECTIVE MEMORY

The politics of memory require a careful negotiation between forgetting and remembrance—they require coming to terms with the presence of a past that is layered and insistent, which refuses to release its hold on the present until its injuries have been addressed. Negotiating the presence of the past involves coming to terms with the complexity of time itself as out of joint—a recognition of time as not governed by teleological norms of progress or universal moral imperatives, but rather, time as nonlinear, and disruptive, where the tenses rub up against each other. It requires conceiving memory not as a "storage model", where pieces of the past are preserved intact for conscious retrieval in the present, but as fragmented and present through multiple layers of perception and recollection. As such, the politics of memory involves a complex process of both tying together and distinguishing fragments of past from the present. It involves periodically renegotiating the boundary between how much to remember and how much to forget.

What the three cases of Algeria's National Charter, South Africa's TRC, and Rwanda's *gacaca* reveal, particularly when considered in relation to each other, is a recognition of the complexity of this negotiation. Each attempted to make sense of the collective fractured memory and accepted the necessity of doing so in the present. None offered an ideal approach to negotiating this thorny terrain, for such a tidy solution is impossible when dealing with the complexity of time, memory, and pain. Thus, when Algeria tried to forget, to release the hold that the past had on the present, through the collective act of voting on a referendum for reconciliation, it risked excluding those who are not able to forget from participating in its political future. South Africa's TRC tried to heal its injuries through acts of truth telling that assume a redemptive past that can transform it, but risked reproducing silences and exclusions and perpetuating a class of victims. Rwanda's *gacaca* tried to heal the wounds of its horrifically injured past, which was marred by a genocide that left few untouched, through an innovative process of community-based justice, that remembered its past in multiple and fragmentary ways. It not only risked reproducing a fragmentation that could bury the past, but it also had considerable strengths in its decentralized and layered approach.

The politics of memory also require acts of overcoming, which harness the plastic power of its people. On this account, it is less clear whether these negotiations of the politics of memory can achieve true reconciliation. It might seem unabashedly romantic when writing of societies that have suffered such grievously injured pasts to ask whether they can overcome their pasts through creative acts in the present that to affirm life—a life that not only is always full of pain and suffering but also contains moments of happiness and joy. But reconciling the presence of an injured past requires such acts—in order to overcome the spirit of existential revenge. The tragic condition of life requires coming to terms with the inescapability of pain and suffering but finding the strength and courage to go forward into the future without fostering resentment. This desire to seek vengeance can incapacitate entire countries. It can lead them into interminable cycles of retribution. While the pain of the past cannot be forgotten, perhaps it can transfigure the will to vengeance and pursue a politics of the future.

On this count, Algeria's blanket legislative act of forgiveness might well help it to move forward with its agenda of reform, which has been stifled by the overwhelming presence of a past that has blocked political progress. However, it seems unlikely that an all-encompassing act that simply turns the page on countless individual memories through public legislation will be sufficient to quell the spirit of revenge in Algeria, where these memories have not been transfigured through affirmative acts but are in some cases barely acknowledged and repressed. In South Africa, the TRC established a comprehensive system to create a national collective memory of its apartheid past and hoped to achieve reconciliation through numerous redemptive acts of truth telling. South Africa seems to have avoided vengeful acts of retribution, but in its universalist approach to the fragmentary nature of memory, it also silenced historically marginalized constituencies and shielded powerful interests from the light of memory. It is also unclear whether the many acts of forgiveness exhibited during the TRC process will be enough to foster reconciliation, whether the very act of remembrance itself is sufficient to redeem the past. Perhaps Rwanda's *gacaca* process comes closest to approximating Nietzsche's idea of a community exercising its plastic power to relinquish its resentment against past suffering by attempting to transfigure it through creative acts in the present. As one part of a tripartite response to the horrors of its genocide, *gacaca* allowed members of the community to participate in proceedings against those who committed grievous crimes against them. In these hearings, members of the community sometimes coalesced multiple memory-fragments into a collective account; in other instances, they were jarred by accusations of complicity in crimes, which viscerally illustrated the disruptive nature of the past; and sometimes they memorialized the death of loved ones through individual and collective acts of remembrance that held out the possibility of healing. In negotiating the politics of memory in a manner that was culturally specific, that allowed for informal participation by local communities, and that was as fragmentary and dissonant as memory itself, where acts of reconciliation remained a constant struggle, Rwanda's *gacaca* perhaps came closest to coping with the tragic nature of its past.

In a world that is replete with cultural difference, where it is no longer possible to pose a comprehensive account of "our time" as Sheldon Wolin has suggested, where there are multiple culturally constituted concepts of time and where different societies remember and forget their pasts in different ways, posing a comprehensive and universal solution to the politics of memory is impossible. In a world of complex time, where the tenses shove and shape each other, where the past is always present and the future already impinges on the now, the politics of memory must remain multifaceted enterprise. Negotiating the complex currents of time and memory always involves asking how much of the past to remember and how much to forget and how to overcome the spirit of revenge so readily imposed by the past against the layered dispositions of the present. It remains an open question as to whether the horrible traumas from the past can be transfigured into conditions of possibility for a better future.

NOTES

1. William Faulkner, *Requiem for a Nun* (New York: Vintage Books, 1975).
2. Michel-Rolph Trouillot, *Silencing the Past: Power and the Production of History* (Boston: Beacon Press, 1995), p. 16.
3. Gilles Deleuze, *Cinema 2: The Time-Image*, trans. Hugh Tomlinson and Robert Galeta (Minneapolis: University of Minnesota Press,1997), p. 99.
4. Henri Bergson, *Matter and Memory*, trans. N.M. Paul and W.S. Palmer (New York: Zone Books, 2002), p. 35.
5. Trouillot, *Silencing the Past*, p. 16.
6. Michael Ignatieff, "The Elusive Goal of War Trials", *Harper's Magazine*, vol. 294, no. 1762 (March 1997): 15.
7. Friedrich Nietzsche, *Untimely Meditations*, trans. R.J. Hollingdale (Cambridge: Cambridge Universtiy Press, 1997), p. 61.
8. Ibid., p. 61.
9. Ibid., p. 62.
10. Ibid., p. 62.
11. Alix Spiegel, "Can We Learn to Forget Our Memories?", *National Public Radio*, September 3, 2012, www.npr.org/blogs/health/2012/09/03/160126283/can-we-learn-to-forget-our-memories.

12. Nietzsche, *Untimely Meditations*, p. 62.
13. Ibid., p. 61.
14. Ibid., p. 62.
15. Ibid., p. 63.
16. Ibid., p. 63.
17. Martha Minow, *Between Vengeance and Forgiveness: Facing History after Genocide and Mass Violence* (Boston, MA: Beacon Press, 1998), p. 2.
18. Sheldon Wolin, *What Time is it?* in *Theory and Event* 1:1 (Project Muse, Johns Hopkins University Press, http://muse.jhu.edu/journals/tae, 1997).
19. Ibid., p. 5.
20. Ibid., p. 4.
21. Michael Slackman, "Top Algerians Prefer Amnesia to Accountability on War", *The New York Times*, September 26, 2005.
22. Ibid.
23. "A Flawed Charter", *The Economist* 337, no. 8447, October 8, 2005.
24. Slackman, "Top Algerians."
25. "Algeria Referendum Numbers Queried", *Africa Research Bulletin: Political Social and Cultural Series* 42, no. 10 (October 1–31, 2005): 16383–16384.
26. "A Flawed Charter."
27. "Aims and Objectives of the Charter", *Algeria Today*, September 30, 2005, www.algeria-us.org/ALGERIA_TODAY_0%5b1%5d.pdf.
28. Ibid.
29. Slackman, "Top Algerians."
30. Ibid.
31. Ibid.
32. Michael Slackman, "But Bygones Can't Be Bygones if the Pain Is Raw", *The New York Times*, October 5, 2005.
33. Ibid.
34. George Joffé, "National Reconciliation and General Amnesty in Algeria", *Mediterranean Politics* 13, no. 2 (July 2008): 222.
35. Ibid., p. 223.
36. Pieter Duvenage, "The Politics of Memory and Forgetting after Auschwitz and Apartheid", *Philosophy and Social Criticism* 25 no. 3 (1999): 13.
37. James L. Gibson, "Overcoming Apartheid: Can Truth Reconcile a Divided Nation?" *Politikon: South African Journal of Political Studies* 31 (November 2004): 130.
38. The Promotion of National Unity and Reconciliation Act of 1995 (Act. 95-34, July 26, 1995), www.justice.gov.za/legislation/acts/1995-034.pdf, p. 1.
39. Ibid., p. 6.
40. Ibid., p. 6.
41. Gibson, "Overcoming Apartheid," p. 132.
42. Minow, *Between Vengeance and Forgiveness*, p. 61.
43. Ibid., p. 70.
44. Gibson, "Overcoming Apartheid", p. 139.
45. Minow, *Between Vengeance and Forgiveness*, p. 84.
46. Trouillot, *Silencing the Past*, p. 27.
47. Michel Foucault, "Truth/Power", in *The Foucault Reader*, ed. Paul Rabinow (New York: Random House, 1984) p. 74.
48. Minow, *Between Vengeance and Forgiveness*, p. 75.
49. Cheryl McEwan, "Building a Postcolonial Archive? Gender, Collective Memory and Citizenship in Post-apartheid South Africa", *Journal of Southern African Studies* 29, no. 3 (September 2003): p. 741.
50. Ibid., p. 745.
51. Ibid., p. 745.

52. Ibid., p. 746.
53. Samantha Power, *A Problem from Hell: America and the Age of Genocide* (New York: Perennial Books, 2003) p. 334.
54. Jason Strain and Elizabeth Keyes, "Accountability for Atrocities in the Aftermath of Rwanda's Genocide" in *Accountability for Atrocities: National and International Responses*, ed. Jane E. Stromseth, (Ardsley, NY: Transnational Publishers, Inc, 2003), p. 94.
55. Ibid., p. 94.
56. Ibid., p. 97.
57. Ibid., p. 106.
58. Ibid., p. 109.
59. Ibid., p. 110.
60. Philip Clark, "When the Killers Go Home: Local Justice in Rwanda", *Dissent Magazine* (Summer 2005), p. 14.
61. Strain and Keyes, "Accountability for Atrocities", p. 116.
62. Clark, "When the Killers Go Home", p. 14.
63. Ibid., p. 14.
64. Philip Clark, "The Legacy of Rwanda's Gacaca Courts", *Think Africa Press*, March 23, 2012, http://thinkafricapress.com/rwanda/legacy-gacaca-courts-genocide.
65. Strain and Keyes, "Accountability for Atrocities", p. 118.
66. Ibid., p. 118.
67. Clark, "When the Killers Go Home", p. 21.
68. Ibid., p. 20.
69. Strain and Keyes, "Accountability for Atrocities", p. 119.
70. Ibid., p. 121.
71. Ibid., p. 121.
72. Ibid., p. 121.
73. Clark, "When the Killers Go Home", p. 19.
74. Ibid., p. 20.
75. Ibid., p. 20.
76. Ibid., p. 21.
77. Ibid., p. 20.
78. Strain and Keyes, "Accountability for Atrocities", p. 131.
79. Philip Clark, *The Gacaca Courts, Post-Genocide Justice and Reconciliation in Rwanda: Justice without Lawyers* (Cambridge: Cambridge University Press, 2010), pp. 352–354.
80. "NURC, Opinion Survey on Participation in Gacaca", Annexe 3m p. 9, in Clark, *The Gacaca Courts*, p. 264.
81. "NURC, Opinion Survey on Participation in Gacaca", Annexe 3, p. 18, in Clark, *The Gacaca Courts*, p. 299.
82. "NURC, Opinion Survey on Participation in Gacaca", Annexe 4, p. 10, in Clark, *The Gacaca Courts*, p. 299.
83. National Unity and Reconciliation Commission (NURC), *Social Cohesion in Rwanda, An Opinion Survey, Results 2005-2007*, www.nurc.gov.rw/index.php?id=69&eID=dam_frontend_push&docID=12, p. 5.
84. Ibid., p. 5.
85. Clark, *The Gacaca Courts*, p. 354.

5 Secular Time and the Politics of Renewal

In recent years, secularism as a political doctrine has come under scrutiny from various angles. Its universal ambitions have been questioned and its connection to modernity with which it is inextricably linked has been called up for interrogation. It is important to note at the outset that the separation between religion and politics, having emerged out of a particular Western context of religious conflict, in which the political was shaped by its appropriation and deployment of religious doctrine, has never been as airtight as many of its proponents would suggest. Late modern critics of secularism who seek a more pluralizing form of it have argued that "the secular modus vivendi ignores or devalues some dimensions of being that need to be engaged more openly".[1] This line of argument suggests that by limiting public discourse to rational deliberation, secularism ignores the visceral dimension of thought. By pushing such elements to the margins, secularist politics has drained the public sphere of much needed vitality and diversity by demarcating it as a space of cold rational calculation. Indeed, secularism goes further than simply separating religion from politics and emotion and affect from rational deliberation. As Saba Mahmood has suggested, "secularism has sought not so much to banish religion from the public domain but to reshape the form it takes, the subjectivities it endorses, and the epistemological claims it can make".[2]

In this chapter, I attempt to examine the "reshaping" effect of secularism from a temporal perspective. I argue that the time-image undergirding secularism as a political doctrine is inextricably linked with the question of modernity's time consciousness, which is often articulated as emerging when "Christian eschatology shed its constant expectation of the imminent arrival of doomsday", which allowed for a time-image to "be revealed that would be open for the new and without limit".[3] This modern, and explicitly secular, time-consciousness suggested that history could be regarded as a "long-term process of growing fulfillment which, despite setbacks and deviations, was ultimately planned and carried out by men themselves".[4] This chapter critically explores the question of secular time, analyzes its emergence and construction, and explores the limits of its application to a world of difference where time is experienced as out of joint. In particular, it asks to what extent can a homogeneous and sequential time-image that emerges out of a particular Christian and European context be applied to the Islamic world and, specifically, to a strain of Islamic fundamentalism that emphasizes renewal? In exploring this question, it is necessary to note that I am less interested here in making the case that Islam is interpellated in and participates in the discourse of modernity, or in the useful conceptual language of "multiple modernities", as much as I am focusing on acknowledging modernity as a political reality and project that has to be engaged. In doing so, I follow Talal Asad, who has written that "[t]he important question, therefore, is not to determine why the idea of 'modernity' (or 'the West') is a misdescription, but why it has become hegemonic *as a political* goal . . . It is right to say that 'modernity' is not totally a coherent object nor a clearly bounded one . . . Modernity is a *project*—or rather, a series of interlinked projects—that certain people in power seek to achieve".[5] As such, I argue that Islamic perspectives can be seen as a particular discursive response from a specific cultural context to the political project of modernity.

In order to explore the image of time that structures secularism as a political doctrine, I begin by examining Charles Taylor's image of secular time as "profane" time, as sequential, structured by progress, and emptied of its metaphysical connection to the divine, to ask to what extent is it possible to speak of the secular age, of a homogeneous experience of time in which we can universally participate? I then unpack this question in the context of the contemporary discourse of the West's relationship with the Islamic world. In recent years, the Islamic world has been characterized as a monolithic whole, that is antimodern by dint of what is perceived to be a broad rejection of secularism and that yearns for a fossilized past and remains hopelessly anachronistic in its refusal or inability to experience and inhabit secular time. In order to explore this temporal tension, I examine the writings of Sayyid Qutb, who was not only shaped by the complex legacy of Egypt's colonial experience but who also wrote his famous text *Milestones*, at least in part, as a response to the modernizing and secularizing initiatives undertaken by Nasser. Qutb provides a complex temporal perspective by arguing for a renewal of the Islamic tradition and by mounting a critique of the idea of progress that sustains secular time, insisting instead on the revival of the past to energize the present in order to open up the possibility of a utopian future. By reading Qutb alongside Taylor's account of homogeneous secular time, I aim to explore the insufficiency of such an image of time for a globalized world marked by shifting relations of contingency, where time is experienced as disruptive, heterogeneous, and out of joint.

FROM THE SACRED TO THE PROFANE: THE EMERGENCE OF SECULAR TIME

Secularism, as a political and philosophical doctrine, has always carried within it a temporal dimension. The word *secularism* itself is derived from a Latin term that was originally employed in a Christian context, and interestingly, the origin of the term *secular* has from its earliest use contained within it a particular time image. In *Modes of Secularism*, Charles Taylor explains that the term *secularism* is derived from the word *saeculum*, which was "the ordinary Latin word for century, or age", which later "took on a special meaning as applied to profane time, the time of ordinary historical succession, which the human race lives through between the Fall and the Parousia".[6] Secularism therefore carried within it a profound temporal dimension. From its origins, it was a term that denoted an epoch or period that was conscious of its separation from the past, and perhaps its own historical significance. As a term signifying a particular "age", it already contained within it an element of temporal distinction or demarcation. This was further heightened by the transformation of its meaning into a profane time, a time that was stripped of religious significance.

Such a "profane" image of time could therefore be distinguished from the Christian experience of time that had a defined beginning and a determined end and could be understood as the progressive realization of salvation through the grace of God. This higher or sacred time was therefore enmeshed with profane or secular time and human beings participated in both. As Taylor writes, "This time was interwoven with higher times, different modes of what is sometimes called 'eternity', the time of the Ideas, or the origin, or of God. Human beings were seen as living in all these times, but certain acts, or lives or institutions, or social forms could be seen as more thoroughly directed towards one or another. Government was more 'in the *saeculum*' by contrast with the Church, for instance. The state was the 'secular arm.'"[7] Taylor suggests that there was a duality of temporal perspectives—one could inhabit Christian time, seen as the progressive realization of salvation, whereas the political could still take place in the profane time that was stripped of religious significance and emptied of its mediating content. Thus, while both images of time were linear and sequential, one was imbued with religious significance while the other was seen simply as chronological succession where the everyday bureaucratic life of the state was experienced. What is interesting about this temporal division is not only the emptying of religious meaning from temporality but also the idea of a different time for politics, where the state and its operations are placed in a different temporal register from the outset.

Secularism is therefore not merely a doctrine that separates religion from politics into two different spheres of the public and private, but it is also one that from its origins assigns them to two different images of time that weave in and out of each other. As Taylor writes in *The Secular Age,* the sacred, higher times "gathered, assembled, reordered, punctuated profane, ordinary time".[8] The time of the secular is therefore both separate from, and experienced alongside, the "higher" time of the sacred. As a consequence of this separation, it is "punctuated" by Christian time but is not imbued with the significance of the latter. It is assigned to "this" world and not the transcendental register of the divine. Taylor explains that the

> existence of these oppositions reflected something fundamental about Christendom, a requirement of distance, of non-coincidence between the Church and the world. There were through the medieval centuries great overlap and great conflict between Church and state, but in all versions and on all sides, it was axiomatic that there had to be a separation of spheres . . . more fundamentally, the need for distance, for a less than full embedding in the secular, was understood as essential to the vocation of the Church. One of the motivations for defining a space of the secular has always been theological in Christendom and continues to be so today.[9]

There was a theological distinction between the spiritual and the profane realm at the heart of Christianity that led to a cleaving of temporal perspectives, with one stripped of all religious content and smoothed into homogeneity. Interestingly, this division only gives us a sense of the contemporaneity of different temporal perspectives. However they might be demarcated, as higher and lower, as sacred and profane, they existed in a kind of simultaneity. One could be Christian and have a linear, theologically understood, and enacted temporal perspective, but could simultaneously switch to secular time for profane tasks.

Secular time underwent subsequent transformations after its initial delineation as the successive time of the political to reach its status as ordinary time, the time of the everyday. The simultaneity of the earlier temporal demarcation gave way to the dominance of secular time, during the Enlightenment, an age in which Christian faith was called into question and found lacking and the notion of a sacred or higher time as progressive realization of redemption that had a defined end of God's choosing was less tenable. As Georgio Agamben writes, "The modern concept of time is a secularization of rectilinear, irreversible Christian time, albeit sundered from any notion of end and emptied of any other meaning but that of a structured process of before and after".[10] Thus, while secular time retained the linear form bequeathed to it by Christian theology, it stripped it of its finalism and eventually took its place as the dominant conception of time, which was no longer shaped and interwoven with sacred time. The continued secularization of time and its emergence as a dominant model also transformed the way in which it viewed the future. It contained a crucial new temporal relationship between the present and the future, which when liberated of the religious frame of an imminent doomsday allowed for a more open conception of the future as a source of endless potential and progress. Not only was the future emptied of its religious significance; it also was given new meaning as the focal point of temporal experience and eagerly anticipated as a source of change and as a limitless horizon of expectation. Time, stripped of its eschatological meaning, was now "no more than a simple succession of now in terms of before and after".[11] In Agamben's terms, what was previously a rich record of the history of salvation had become "pure chronology", and consequently, "a semblance of meaning [could] be saved only by introducing the idea . . . of a continuous, infinite progress".[12]

THE HOMOGENOUS TIME OF THE SECULAR

The secular transformation of temporality therefore detached itself from the "higher" shaping of Christian time but it also took on an additional political dimension. The secular, as I have shown earlier, was seen from the beginning as the temporal register of political activity, but it was now transformed into the time we all share. According to Taylor, "the inescapability of secularism flows from the nature of the modern state".[13] Taylor follows Benedict Anderson, who argues in *Imagined Communities* that one of the defining features of the nation-state is its common experience of homogeneous time, which is "a secular transformation of fatality into continuity, contingency into meaning".[14] Thus, according to Taylor, an "important feature of the modern social imaginary is that it no longer sees the greater translocal entities as grounded in something other, something higher, than common action in secular time".[15] He sees the American Revolution as a watershed in this new temporal orientation that requires new higher grounding than common experience and action of a people. He writes, "here the idea is involved that a people, or as it was also called at the time, a 'nation', can exist prior to and independently of its political Constitution, so that this people can give itself its own Constitution by its own free action in secular time"[16]. Taylor further argues that the common experience of secular time in the nation-state, as one that we all experience and participate in is a consequence of its simultaneity. However, simultaneity in his argument means something other than the interweaving of temporalities that characterized the early division of sacred and profane time. Secular time, stripped off its religious significance, is now mere succession and simultaneity and now implies the simultaneous experience by members of the nation-state of the same instant of chronological time. Taylor writes:

> the new sense of belonging to a nation was prepared by a new way of grasping society under the category of simultaneity: society as the whole consisting of the simultaneous happening of all the myriad events that mark the lives of its members at that moment. These events are the fillers of this segment of a kind of homogenous time. This very clear, unambiguous concept of simultaneity belongs to an understanding of time as exclusively secular. As long as secular time is interwoven with various kinds of higher time, there is no guarantee that all events can be placed in unambiguous relations of simultaneity and succession. The high feast is in one way contemporaneous with my life and that of my fellow pilgrims, but in another way it is close to eternity, or the time of origins, or the events it prefigures.[17]

But to what extent is it possible for members of a nation to simultaneously experience secular time? The simultaneity that Taylor and Anderson espouse is actually the simultaneity of multiple complex temporal perspectives, which calls into question its assertion of homogeneity as the basis for community formation. Taylor suggests that such an account is "unambiguous", but instead it turns out to be full of ambiguity and complexity. The argument is that each chronological moment is no longer connected to higher or sacred time and therefore can provide a unifying foundation. Taylor writes that "a purely secular time-understanding allows us to imagine society horizontally, unrelated to any 'high points', where the ordinary sequence of events touches higher time, and therefore without recognizing any privileged persons or agencies—such as kings or priests—who stand and mediate at such alleged points".[18] What this seemingly democratic and smoothing perspective ignores is the complexity of the multiple experiences of a chronological instant. Even if secular time is stripped of its "high points" of religious significance, every moment of time still connects to different points for different people in different pasts that seek to locate it in different futures. If a segment of secular time is made up of all the members experiencing the same instant, at that very moment, there exists a multiplicity of temporal perspectives, multiple pasts that are being called on to and jostled into the present, multiple appropriations of the future that are pulled up by a diversity of expectations that impact the present. Indeed, that instant is layered and shot through with the complexity of numerous experiences of time as out of joint.

The simultaneous instant is not experienced in the same manner by every member of a political community and if we attend to the contingent and destabilizing experience of time as out of joint, it becomes clear that it gives rise to temporal experiences that are precisely not homogenizing. We do not all imbue the same moment with the same significance and, in fact, we may not agree on the significance of the same present itself. Consider, for instance, a political speech, a State of the Union address or something similar, which is considered to be a unifying moment politically. Is that instant experienced simultaneously by all of us? It may occur at a common moment of chronological time but it is imbued with significance in multiple ways and no less so than if it was linked to higher time. One might not watch or listen to the speech and even be unaware of its happening, one might intentionally avoid it as an act of political refusal, one might watch it with a memory of a previous speech and note the ways in which it departs from or confirms our ideas of the imagined community in which we live, one might not consider it to be significant at all depending on the degree of our inclusion or exclusion in the community itself. Secular time then appears to be much less homogeneous than Taylor argues it is, and instead emerges as a heterogeneous experience of time in which multiple pasts are called into the present for the sake of multiple and diverging futures.

Taylor's rendering of secular time as homogenizing draws heavily on Benedict Anderson's argument about the temporality of the nation, which in turn is based on a reading of Walter Benjamin's famous account of messianic time in *Theses on the Philosophy of History.* Anderson's discussion of homogenous time rests on a narrow reading of Walter Benjamin's discussion of "now-time". Anderson writes that "what has come to take the place of the medieval simultaneity-along-time is, to borrow again from Benjamin, an idea of 'homogeneous, empty time' in which simultaneity is, as it were, transverse, cross-time, marked not by prefiguring and fulfillment, but by temporal coincidence, and measured by clock and calendar".[19] While Anderson takes from Benjamin the concept of homogeneous time, he leaves out Benjamin's central account of the disruption of homogeneous time such that "history is the subject of a structure whose site is not homogeneous, empty time, but time filled by the presence of the now".[20] As Benjamin writes, "the true picture of the past flits by. The past can be seized only as an image which flashes up at the instant when it can be recognized and is never seen again . . . For every image of the past that is not recognized by the present as one of its own concerns threatens to disappear irretrievably".[21] For Benjamin, redemption in the future requires the capacity to possess a sliver of the past, to seize it in the present in an out of joint moment of uncertainty in order to transfigure it and create new possibilities for the future. Anderson, however, writes that such an idea "of *simultaneity* is wholly alien to our own. It views time as something close to what Benjamin calls Messianic time, a simultaneity of past and future in an instantaneous present. In such a view of things, the word 'meanwhile' cannot be of real significance".[22] For Anderson, the secular and homogenizing time of the nation-state can be simultaneous in its contemporaneity, but it cannot be simultaneous in the sense of the disruptive and multiple experiences of time that Benjamin articulates. Anderson therefore makes use of the category of homogeneous time that Benjamin discusses while stripping from it its key disruptive quality in which the new emerges in a moment that is wholly out of joint, in which the past is called into the present to redeem the future and splinters homogeneous time into heterogeneity in an instant. In eliminating from secular time this profoundly complex experience of simultaneity, Anderson renders the homogeneous time of the state into a linear chronological sequence that purports to universality and must rely on the idea of progress to structure it.

Even as he follows Anderson in his analysis, Taylor comes to what seems to be a partial recognition of the insufficiency of a common secular and homogeneous experience of time when he points to the possibility disenchantment it can carry within it. He writes that

> our encasing in secular time is also something we have brought about in the way we live and order our lives. It has been brought about by the same social and ideological changes which have wrought disenchantment. In particular, the disciplines of our modern civilized order have

> led us to measure and organize time as never before in human history. Time has become a precious resource, not to be "wasted". The result has been the creation of a tight, ordered time environment. This has enveloped us, until it comes to seem like nature. We have constructed an environment in which we live a uniform, univocal secular time, which we try to measure and control to get things done.[23]

While Taylor points to the acceleration, loss, and progression of secular time, he does not call into question its assertion of common temporal experience, referring instead to "our encasing", "our modern", and a "uniform, univocal secular time" that "we" all attempt to order. As Talal Asad points out, "although it is subject, like all remembered time, to continuous re-formation, reinvestment, and reinvocation, secular history's linear temporality has become the privileged measure of all time".[24] The linear and homogeneous account of secular time is now described as dominant, but what experiences are lost in the assertion of its privilege, and can such an assertion be sustained in a complex and interdependent world of difference?

SECULAR TIME, ANACHRONISM, AND POWER

In a fascinating essay on sexual politics and secular time, Judith Butler takes up this question of the cultural constitution of time and extends it further with respect to debates about Islam, secularism, and modernity. Butler writes that such debates are "already imbued with the problem of time, of progress in particular, and in certain notions of what it means to unfold a future of freedom in time. That there is no one time, that the question of what time this is, already divides us, has to do with which histories have turned out to be formative, how they intersect—or fail to intersect with other histories—and so with a question of how temporality is organized along spatial lines".[25] The articulation of secular time as homogeneous, linear, and structured by an understanding of progress is therefore problematic on at least two levels. First, it does not account for the complex lived experience of time as out of joint, in which the past, present, and future shove and shape each other, and assumes the commonality of lived time, which is cast as homogeneous once it is emptied of its religious significance. Second, it assumes the universal application of secular time, giving short shrift to the cultural constitution of time and the plethora of traditions and cultural discourses that give shape to temporal experience. In doing so, it sharply narrows the temporality of the political as forever accelerating, fleeting, and determined by progress and leaves little room for an emphasis on tradition, memory, or individual experience. Butler complicates the narrative of secular time as she writes that "the problem is not that there are different cultures at war with one another, or that there are different modalities of time . . . that are articulated in different and differentiated cultural locations or that come into confused or brutal contact with one another" but that such a dominant account misses an important point, namely, "that hegemonic conceptions of progress define themselves over and against a pre-modern temporality that they produce for the purposes of their own self-definition. Politically, the questions, what time are we in? are all of us in the same time? and specifically, who has arrived at modernity and who has not? are all raised in the midst of very serious political contestation".[26]

Secular time, as argued by Taylor and Anderson treat these categories as fixed and settled, with their assumption of time as linear and homogeneous, not as disruptive and dislodging. Such an assertion of commonality and homogeneity, which is inextricably linked with the discourse and political reality of modernity, necessarily finds it difficult to reckon with different perspectives that do not fit into its narrative. Much ink has been spilled in recent years by authors who point to Islam, in particular, as a culture that has not achieved modernity, because of its purported resistance to secularism.[27] As such, Islam (usually treated as a monolithic whole) is cast as being stuck in a fossilized past, as somehow outside of secular time, and as a temporal oddity that unsettles its universal narrative. As Butler puts it, "anachronistic Islam is figured here as the child who suffers permanently from thwarted development".[28] Key to the narrative of secular time, once it is emptied of religious significance, is the idea of progress as a regulative idea and Islam is cast as unable to achieve its requisite place in this modernizing narrative. What this ignores, of course, is the power that is exercised through such a narrative. In Butler's words,

> the problem is, of course, not progress per se, nor surely the future, but specific developmental narratives in which certain exclusionary and ever persecutory norms become at once the precondition and teleology of culture . . . Outside of its own teleology exists a ruinous and foreboding sense of the future, and what lies before its transcendental condition lurks an aberrant anachronism, threatening and intruding upon, the political present that becomes the grounds for general alarm within the secular frame.[29]

The political implications of such a temporal casting are profound. If Islam is cast as an anachronism, as somehow aberrantly outside the secular rendering of "our time", not only does it reveal tensions and cracks within that time image, but it also permits the characterization of Muslims as somehow not modern and not part of the community that is purportedly created on its basis. Butler is explicit about the political dangers that are folded within such an exclusionary narrative when she suggests that

> if the Islamic populations destroyed in the recent and current war are considered less than human or "outside" the cultural conditions for the emergence of the human, than they belong either to a time of cultural

> infancy or to a time that is outside time as we know it. In both cases, they are regarded as not yet having arrived at the idea of the rational; human . . . It is also precisely this particular conceding of a progressive history that positions the "West" as articulating the paradigmatic principles of the human—humans worth valuing, whose lives are worth safeguarding, protecting, whose lives are precarious, and worth public grieving as well.[30]

Such a rendering of the time-image is therefore both insufficient and exclusionary. A universalizing and homogeneous assumption of secular time makes it much more difficult to open conceptual space to explore the experience of time from an Islamic perspective, but also underlines the necessity of doing so in a world of difference.

Talal Asad makes this point clear when he explores the temporal dissonance between these perspectives when he writes that "the temporalities of many tradition-rooted practices (that is, the time each embodied practice requires to complete and to perfect itself, the past into which it reaches, that it reencounters, reimagines and extends) cannot be translated into the homogenous time of national politics".[31] The dissonance, however, does not suggest incommensurability, but rather the necessity of crafting a complex temporal perspective that remains open to these differing temporalities and the tensions that emerge when they intersect with each other. Thus, while secularism and modernity are inextricably linked and emerge from a particular European context of religious conflict that make it difficult to translate into other cultural contexts, nevertheless, modernity exists as a political reality and secular time occupies a place of exclusionary privilege to which other cultures must necessarily respond.

SAYYID QUTB AND THE IDEA OF RENEWAL

The insufficiency of secular time and the exclusionary power it has folded within it makes it important to engage a different perspective, or what Roxanne Euben has termed "a view from another side".[32] In order to do so, I turn to an exploration of Sayyid Qutb's concept of renewal as it is articulated in his famous and controversial text *Milestones*. Qutb is an engaging and problematic figure who puts forward a trenchant critique of the West as morally bankrupt and having veered off course. Qutb is particularly interesting because he extends his critique to the Islamic world that he sees as infected with the same corruption. This is not to say that *Milestones* is primarily a reactionary text in its response to Western notions of modernity, but instead to point out that as someone who was shaped by Egypt's colonial legacy, deeply influenced by a formative trip to the United States, and caught in the political crosshairs of Nasser's attempts to "modernize" Egypt in a European mold, Qutb's political reality was enmeshed with Western influence. Qutb's scathing critique of the West, and the Islamic world, as infected with moral bankruptcy, is precisely not an anachronistic premodern reaction to secularism, but rather a particularly modern response from within the discourse of Islam. It is an attempt to address the political conditions of the present through the work of critique and in doing so it provides an interesting perspective that needs to be engaged. As Euben writes, "Qutb's Islamic fundamentalism retrieves and reinterprets the lessons embedded in an 'original' Islamic ideal as a guide to radically restructure modern political and social conceptions of the world".[33] I therefore argue that it is precisely not that Qutb's writings operate outside of secular time as a temporal anachronism that should be excluded from the discourse of modernity, but rather that his temporal perspective reveals tensions within the image of secular time that need to be explored.

In the acts of retrieval and reinterpretation, Qutb provides us with an interesting temporal perspective to consider. *Milestones* is fueled by the concept of renewal as essential to the Islamic project—a renewal of an authentic Islamic past, to jostle the present in order to open up the possibility for an enchanted future. As such, it calls into question the strict linearity of homogenizing secular time and opens the door to a more complex understanding of the experience of time that recognizes the fluidity of the tenses. There are fugitive moments in his thinking that leaves a trace of complex temporality. While Qutb ends up recasting the future in a fairly utopian fashion (what Andrew March has termed a "realistic utopia") that is not open to pluralism or the politics of difference, his critique of the idea of progress that structures secular time, and his attempt to complicate its temporal perspective nonetheless remains compelling.

Milestones begins with a premonition of danger and a diagnosis of the modern condition. In its opening lines there is a sense of disenchantment with the modern that is not entirely incompatible with Taylor's analysis, but the tone is sharper and the sense of impending catastrophe makes for a compelling beginning. Qutb writes, "Mankind today is on the brink of a precipice, not because of the danger of complete annihilation which is hanging over its head . . . but because humanity is devoid of those vital values which are necessary not only for its healthy development but also for its real progress" and continues to suggest that not only has progress been derailed because of this moral bankruptcy but that the recognition of this condition is a shared experience, such that "even the Western world realizes that Western civilization is unable to present any healthy values for the guidance of mankind. It knows that it does not possess anything which will satisfy its own conscience and justify its own existence".[34] The modern world for Qutb is a nihilistic one that is unmoored from the structuring foundations of faith. In a sense, Qutb seems to underline the early Christian division between the sacred and the profane world of politics, but he is especially attuned to the damage done by this emptying of the public sphere.

Qutb's prescription for this gloomy diagnosis is to call for a political revival of Islam and restore the link between the profane and the sacred. He rejects the separating premise of secular modernity entirely and in doing so rejects the secular time-image as well. He writes, "If Islam is again to play the role of the leader of mankind, then it is necessary that the Muslim community be renewed to its original form".[35] In a key formulation, he makes clear that the solution to the disenchantment of the modern is to retrieve a heretofore hidden past that has been obscured but still remains present and needs to be actualized in the moment for the sake of the future. He writes that "it is necessary to revive that Muslim community which is buried under the debris of the man-made traditions of several generations and which is crushed under the weight of those false laws and customs which are not even remotely related to the Islamic teachings, and which, in spite of all this, calls itself the 'world of Islam.'"[36] The response to the disenchanted present therefore requires the renewal of a past, even though such an act of renewal is not easy. Indeed, he is "aware that between the attempt at 'revival' and the attainment of 'leadership' there is a great distance, as the Muslim community has long ago vanished from existence and from observation".[37] The past is distant and obscured, but it is not inaccessible and any political movement that wants to change the present condition must necessarily attempt to revive it and reenergize the slumbering community of Muslims (the *umma*). As Euben explains, "As used by Qutb, the *umma* is thus a transcendent, ahistoric ideal waiting to be actualized at any moment in history . . . In this insistence, Qutb detaches the ideal Muslim community from its historical moorings, investing Islamic texts with new immediacy".[38]

The "new immediacy" that Qutb infuses into the Islamic past operates in tension with the account of secular and homogeneous time we encountered earlier in Taylor and Anderson's writings. First, for Qutb, time is still connected to a traditional or sacred past and such a connection is essential to this modern political project. It is not necessarily linked to points of "higher time", but it is a temporal perspective in which the past of the first generation of Muslims remains present and can be drawn into the present in an out of joint moment by a new vanguard that wants to shape its future. The sacred and the profane are not neatly demarcated as much as they are fully imbricated in a particular cultural experience of time that draws on a living Islamic tradition. Absent the Christian theological impetus to separate the profane from the sacred, the religious, and the political are enmeshed with each other and are part of each other's pasts, presents, and futures. Time is therefore not emptied of religious significance, nor is it clearly structured by it, inasmuch as it is imbued by and inseparable from its own cultural and discursive past. As such the past and present are not linked by pure chronological succession but are folded into each other to provide a more complex account of the lived experience of time.

At the heart of this divergence is also a complex critique of the idea of progress that structures secular time. For Qutb, the West exists not just as a source of nihilism but also as one worthy of admiration, even as its impact must be politically contested. He writes of the Enlightenment that

> this was the era during which Europe's genius created its marvelous works in science, culture, law and material production, due to which mankind has progressed to great heights of creativity and material comfort. It is not easy to find fault with the inventors of such marvelous things, especially since what we call the "world of Islam" is completely devoid of all this beauty. But in spite of all this, it is necessary to revive Islam.[39]

Material progress cannot be denied and indeed the creativity it unleashes is essential to the Islamic political project, but by itself, material progress is necessary but not wholly sufficient. Qutb argues that material progress is obligatory and requires the attention of Muslims, but he notes that something else other than technological and material advancement is needed to renew the bankrupt present. He writes that to

> attain the leadership of mankind, we must have something to offer besides material progress, and this other quality can only be a faith and a way of life which, on the one hand, conserves the benefits of modern science and technology, and on the other, fulfills the basic human needs on the same level of excellence as technology has fulfilled them in the sphere of material comfort. And then this faith and way of life must take a concrete form in a human society—in other words, in a Muslim society.[40]

Progress, therefore, requires not just a material but a moral dimension that can respond to human needs of fulfillment. The secular modern is starved of moral nourishment and the separation of the political sphere from the religious has led to moral bankruptcy which complicates the account of material progress. As Euben underlines, Qutb "argues that Western hegemony has ended, not because it is materially or militarily poor, but because the authority of science produced by the Enlightenment worldview has proven incapable of promoting real progress, that is, moral progress"[41]. In other words, the idea of progress, which structures secular time and provides it with its forward motion, is an emptying concept. While hurtling us forward into what it promises will be a better future, it simultaneously hollows out our present and strips it of vital values.

To respond to this decaying present a political movement must be created that renews its past for the sake of the future. It must do so to counter the forces of *jahiliyyah* that have infected both the West and the Islamic world. In classical Islam, *jahiliyyah* was used to describe humanity's state of ignorance before the revelation of Islam. It was used to describe the pagan world, the world of idolaters, and all the conditions and phenomena that were antithetical to the spirit of Islam. Qutb engineers a remarkable transformation and extension of this term—it is now used to describe a world without religious values and foundations, where sources of morality have been corrupted. He writes that "this *Jahiliyyah* is based on rebellion against God's sovereignty on earth. It transfers to man one of the greatest attributes of God, namely sovereignty, and makes some men lords over others. It . . . takes the form of claiming the right to create values, to legislate rules of collective behavior, and to choose any way of life rests with men, without regard to what God has prescribed".[42] To respond to this condition of decay and this usurpation of divine sovereignty by men, "it is necessary that there should be a vanguard which sets out with this determination and then keeps walking on the path, marching through the vast ocean of *Jahiliyyah* which has encompassed the entire world. During its course, it should keep itself somewhat aloof from this all-encompassing *Jahiliyyah* and should also keep some ties with it".[43] The vanguard must therefore separate itself from the *jahili* world but not detach itself completely from it and the material progress associated with it. But to actualize a new political future for Islam, it "must return to that pure source from which those people derived their guidance, the source which is free from any mixing or pollution".[44] The aim is "first to change ourselves so that we may later change the society".[45] *Milestones* is written for this vanguard, which Qutb considers "to be a waiting reality about to be materialized".[46] The text therefore provides a political prescription for a time to come and a time that is almost here. It is written for a political movement that is filled with potentiality and one that works to enact change first on a personal and then on a collective level. It calls for an Islamic political movement structured by a complex temporality, in which an authentic and hidden past is actively worked on in the present and drawn into it by political action to open up the possibility of a new and vital future, free of the corruption of *jahiliyyah*.

It is important to underline that it is not that Qutb has an Islamic image of time that is entirely different from secular time; rather, his account of the experience of time in which a hidden past can be made present brings to light the insufficiency of the linear and homogeneous narrative of the secular time image in which the experience of time is reduced to pure chronology and in which each moment slips from our grasp, never to be recaptured again. Qutb's account reveals the inadequacy of such an analytic framework for coming to terms with a political movement that emphasizes the out of joint experience of time, when an authentic past is brought into the present to renew the future. In doing so, it exposes the cultural specificity of secular time but also calls our attention to the necessity of acknowledging and responding to its limits as a political reality. Qutb's complex temporal perspective is therefore a useful contribution to the debates about secularism and modernity that cannot merely be dismissed as an anachronistic exception to secular time but one that serves an essential critical role in highlighting the insufficiency of its analytic framework.

Secularism as a political doctrine therefore carries within it a particular time-image, which emerged from a Christian theological tradition that demarcated a division between the time of the sacred and the time of the profane. It articulated a concept of time that became the time of the political—a homogeneous, sequential and linear concept of time, structured by progress, and moving unidrectionally toward an open and limitless future. Such an account of time, while emerging out of a specific European historical context of religious conflict, has been cast as the dominant framework, as "our time", as the basis for imagined communities we can all participate in. Such an image of time, characterized as pure chronology, pays scant attention to the complex lived experience of time as out of joint, in which the tenses fluidly intermingle, the past shoves and shapes the now, and the future impinges on the present and molds it. Such an account of temporality also conceals within itself an exclusionary impulse and both defines itself against and has to construct "premodern" concepts of time, to account for temporal perspectives that are culturally and discursively different from its own. Reading the account of secular time as articulated by Taylor and Anderson alongside Qutb's account of renewal in political Islam exposes the insufficiency of secular time for a complex world of difference. Qutb articulates a complex temporal perspective by calling for a renewal of an Islamic past to transfigure a disenchanted present for the possibility of a utopian future. While the future he calls for is one that is resistant to pluralism and difference, it is valuable for dislodging and critiquing key assumptions about the linearity and universality of secular time and provides a valuable perspective from another side. As Talal Asad has written, "in addition to complex space we need to think also of heterogeneous time: of embodied practices rooted in multiple traditions, of the differences between the horizons of expectation and spaces of experience—differences that continually dislocate the present from the past, the world experienced from the world anticipated, and call for their revision and reconnection. These simultaneous temporalities embrace both individuals and groups in complexities that imply more than a process of secular time".[47] By examining temporality from a heterogeneous perspective, we might move closer to such an open and contingent understanding of politics in a world of difference.

NOTES

1. William E. Connolly, *Why I Am Not a Secularist* (Minneapolis: University of Minnesota Press, 2000), p. 19.
2. Saba Mahmood, "Secularism, Hermeneutics and Empire: The Politics of Islamic Reformation", *Public Culture* 18, no. 2 (2006): 326.
3. Reinhart Koselleck, *Futures Past: On the Semantics of Historical Time* (New York: Columbia University Press, 2004), p. 223.
4. Ibid., p. 266.
5. Talal Asad, *Formations of the Secular: Christianity, Islam, Modernity* (Stanford, CA: Stanford University Press, 2003), p. 13.

6. Charles Taylor, "Modes of Secularism", in *Secularism and its Critics* ed. by Rajeev Bhargava (Delhi, India: Oxford University Press, 2005), pp. 31–32.
7. Ibid., p. 32.
8. Charles Taylor, *A Secular Age* (Cambridge, MA: Belknap/Harvard University Press, 2007), p. 54.
9. Taylor, "Modes of Secularism", p. 32.
10. Georgio Agamben, "Time and History—Critique of the Instant and the Continuum", in *Infancy and History: Essays on the Destruction of Experience*, trans. Liz Heron (New York: Verso Books, 1993), p. 96.
11. Ibid., p. 97.
12. Ibid.
13. Taylor, "Modes of Secularism", p. 38.
14. Benedict Anderson, *Imagined Communities: Reflections on the Origin and Spread of Nationalism* (London: Verso, 1991), p. 11.
15. Taylor, "Modes of Secularism", p. 40.
16. Ibid., p. 41.
17. Ibid., pp. 41–42.
18. Ibid., p. 42.
19. Anderson, *Imagined Communities*, p. 24.
20. Walter Benjamin, *Illuminations*, ed. Hannah Arendt and trans. Harry Zohn (New York: Schocken Books, 1968), p. 261.
21. Ibid., p. 255.
22. Anderson, *Imagined Communities*, p. 24.
23. Taylor, *A Secular Age*, p. 59.
24. Asad, *Formations of the Secular*, p. 43.
25. Judith Butler, "Sexual Politics, Torture, and Secular Time", *The British Journal of Sociology* 59, no. 1 (2008): 1.
26. Ibid.
27. See, for instance, the many columns by Thomas Friedman of the *New York Times* or Fareed Zakaria of *Newsweek*, who insist on the necessity of Islam to modernize and participate in the present. Also, see Asad's *Formations of the Secular* on the creation of a "secular" Islamic space from within by liberal reformers such as Mohammed 'Abduh, who make the case for a distinctive Islamic approach to the secular.
28. Butler, *Sexual Politics*, p. 8.
29. Ibid., p. 14.
30. Ibid., p. 15.
31. Asad, *Formations of the Secular*, p. 179.
32. Roxanne L. Euben, *Enemy in the Mirror: Islamic Fundamentalism and the Limits of Modern Rationalism* (Princeton, NJ: Princeton University Press, 1999), p. 49.
33. Ibid., p. 52.
34. Sayyid Qutb, *Milestones* (Damascus: Kazi Books, 1964), p. 7.
35. Ibid., p. 9.
36. Ibid.
37. Ibid.
38. Euben, *Enemy in the Mirror*, p. 61.
39. Qutb, *Milestones*, p. 9.
40. Ibid., p. 10.
41. Euben, *Enemy in the Mirror*, p. 58.
42. Qutb, *Milestones*, p. 10.
43. Ibid., p. 12.
44. Ibid., p. 21.
45. Ibid.
46. Ibid., p. 12.
47. Asad, *Formations of the Secular*, p. 179.

Conclusion

Memories of War and the Politics of Contingency

I was born in the late 1970s in Bangladesh, seven years after the country emerged from a brutal and bloody war for independence from Pakistan. When that protracted conflict finally ended, it gave rise to a promising and even exuberant democratic beginning, which sadly gave way in a few short years to the suspension of fundamental rights, the banning of political parties, the creation of a government sanctioned militia, and profound political mismanagement that led to a devastating and largely preventable famine. The hope and the idealism of the nation's founding, already diminished by the immense losses of the war for independence that left thousands dead and many more the victims of horrific violence, were already struggling to stay alive amidst the miasma of daily despair and the collapse of democratic politics. Three years before I was born, the prime minister and most of his immediate family were massacred in a military coup that was shockingly pitiless in its violence. In its aftermath, the nation lurched from one military dictator, who was assassinated in a subsequent coup, to a series of others, until a democratic transition of sorts took place in 1990.

Growing up in this turbulent era, I understood politics to be marked on ground that constantly shifted and cracked underneath my feet. It was rife with unpredictability—one morning, I came across my mother sobbing in the kitchen because the president, a longtime family friend, had been assassinated; another afternoon, there were reports of tanks in the streets as another short-lived military coup was underway as I sat at my math tutor's house laboring through yet another problem set; another evening, my father and I stood on our rooftop craning to see the helicopter that would signal that yet another president had been removed from office. Political strikes and coups were the snow days of our youth. Even as fourth and fifth graders, my friends and I devoured the political news the way I now consume weather news in Chicago. Instead of trying to figure out if a snowstorm will blanket my morning commute to work as I do on winter mornings now, back in those days my friends and I tried to figure out if the next series of political strikes would mean that our math exam would be pushed back for a week (during one such transition, we were gleeful that finals were postponed for a full month!). But while the moisture in the atmosphere and cold fronts that descend down from Canada are (mostly) beyond our control, necessary, and even predictable to an extent, for me, growing up in Bangladesh, it seemed that the tumult of politics somehow had a similar sense of inevitability, that its very fragility was somehow immutable. We lived in a condition of such impermanence and upheaval that its very contingency started to feel like a permanent feature in our lives, like the snowstorms I now brace for in the Midwest. But unlike the winter weather, which I can mostly count on being behind us in April (or if I'm lucky, March), in Bangladesh we lived with this constant condition of political fragility and contestability that began to feel permanent and enduring. Politics was rife with the ever-present threat of violence and the sense that the present could at any moment be rendered out of joint by another push from the past, from the trauma of war and its memories that insistently refused to leave us, and from the pull of a future that gestured at shaky democratic beginnings that never quite materialized.

In a sense, Bangladesh was, and remains, captive to its violent and turbulent past. Its bloody beginning, which came to an unsatisfying end at the conclusion of a war whose many crimes faded without either investigations or accountability, continues to unsettle the present and cast dark shadows into the country's potential futures. The politics of the present continues to be judged by the allegiances of the past, and war memorialization has been transformed into ideology, used to celebrate allies and tar opponents with a devastating brush, with *muktijoddha* (literally one who fought for freedom) being one of the highest honors one can be referred to by (although who is characterized as such seems to depend not on verifiable military records but on the shifting sentiments of which political party is presently in power) and with *razakar* (which literally translates to *volunteer* and refers to those who sided with and assisted the Pakistani army during the war) pejoratively and often insidiously used to broadly designate the enemies of the state, whose pasts are considered irredeemably treasonous.

In speaking of the war for Bangladeshi independence, which continues to unsettle the present today, even the terms one might use to classify the conflict are more contestable than they might initially appear. After the partition of India, Bangladesh became part of Pakistan, as the other Muslim majority section of India. Referred to as "East Pakistan", it shared a common religion with "West Pakistan", but little else. Culturally and linguistically, Bengalis had far more in common with those of West Bengal, who remained part of India. Pakistani identity was difficult to accept and increasingly contested, as with the Bengali language movement of 1952, which protested the designation of Urdu as the official language of Pakistan and ended in violence, with several of the protestors killed by the police. Power was concentrated in West Pakistan and in the capital of Islamabad. Over the years there were repeated demands for autonomy and a redistribution of power, which came to a head in 1970 when the Bangladeshi leader Sheikh Mujib began a campaign for a Six Point demand for constitutional changes to address those grievances.[1] In December 1970, Mujib and his party emerged victorious in the Pakistani parliamentary elections, winning with an overwhelming 75% of the vote. But the transfer to power never materialized and remained bogged down in endless negotiations. On March 1, the Pakistani president, General Yayha Khan, announced the postponement of the national assembly, which set off furious protests in then East Pakistan. On March 25, 1971, the Pakistani army began "Operation Searchlight", a violent crackdown on the protests that marked the beginning of the war and precipitated nine months of brutal fighting that left thousands of Bangladeshi dead, countless women raped, and large numbers of casualties. But classifying the war is still a matter of dispute—for many in today's Pakistan, the 1971 war is still seen as a "civil war" between warring halves of the same country, whereas in Bangladesh, the "Liberation War", as it is commonly referred to, is cast clearly in the language not just of independence but also of genocide.

The characterization of the war as genocide, contested in some areas, but firmly asserted by the Bangladeshi government and used widely internationally, has led to an unsettling legacy in a country that endured terrible acts of violence but saw no accountability take place for decades for those who committed the war crimes, nor any significant collective attempts to heal from the scars of the past that remain searingly fresh today. One of the biggest problems with the designation of genocide in Bangladesh is that it remains something of a frustratingly empty signifier—it is loaded with the injury and the claims of victimhood that the term represents, but lacks any of the content to realize it in actuality or to give voice to the immense suffering it signifies. When the war finally ended, there were no official investigations, no trials, no serious attempts to account for the dead and injured, and there existed powerful cultural norms of shaming that made accounting for victims of rape almost impossible.[2] The numbers that are often used by both the state and the media of 3 million dead and 300,000 women raped are difficult to verify, and sometimes radically disputed, given this acute lack of accountability.

In recent years, a revisionist look at the Liberation War by the Oxford scholar Sarmila Bose has seriously challenged the veracity of these oft-used numbers that are widely repeated as fact, both domestically and internationally. Bose relied on interviews with Bangladeshi survivors of the war and Pakistani military officials to try to pull together memories of the war, and while her approach runs into several methodological problems and reaches some problematic conclusions, her exploration of the reported violence is significant. She estimates both sets of numbers to be significantly lower than reported and, disconcertingly for many Bangladeshis, also explores shocking accounts of violence by Bangladeshis themselves, against not just the Pakistani armed forces but also against the minority Bihari population in then East Pakistan. Bose's book was widely condemned and criticized in Bangladesh even before it was available there in print and those who spoke to her and assisted her with the project, sometimes welcoming her into their home as a guest, condemned her loudly in the press and spoke of feeling betrayed by her. An op-ed in the most widely circulated English language daily put it in stark terms: "Her book does not change anything. It is wobbly scholarship, a disturbing misreading of history. You lose nothing by not reading it".[3] It is certainly true that Bose's book is problematic on a number of fronts. A former *Guardian* reporter who spent extensive time in Bangladesh during the war, praised her for her attempt to try to knit together an account of the conflict through the compilation of memories but also suggested that it was perhaps too caught up in its revisionist aims. He wrote, "My own feeling, remembering how charming Pakistani officers, like their Indian equivalents, can be, is that she may have been a bit too ready to accept the honourable, just-trying-to-do-our-duty image that those officers naturally prefer to convey, and that she may also be too convinced that the received wisdom needs to be entirely overturned".[4] Methodologically, her approach is problematic, as one reviewer noted, in part because she pays "scant attention to the effect of the passage of time on memories".[5] But to go further, she traffics extensively in the storage model of memory discussed in Chapter 4 of this book by assuming that individual memories are stored intact for conscious retrieval without attending to the how memory is complicated by both perception and recollection and how memories are always shaped by individual experience, disciplined by power relations, and shaped by the particular cultural and political context in which they emerge. Thus, Bose's final assertion that "the 'liberation literature' of Bangladesh repeatedly uses the word 'genocide' . . . in blissful disregard of the need to provide substantiation" and "the persistent cultivation of a 'victim culture' glides effortlessly through allegations of exploitation by West Pakistan, genocide in 1971", needs to be taken with a grain of salt.[6]

Leaving aside the criticism of Bose's methodology, her book and its reception are significant because of the discursive reaction to its publication in Bangladesh. The firestorm that greeted her book signaled that there was little room for alternative accounts or perspectives on the Liberation War in the Bangladeshi political imaginary. It suggested a certain rigidity and a shackling to a particular narrative of the past, which while never fully investigated or accounted for, still has a powerful hold and remains urgently and disorienting present. As unsettling and unstable as the Bangladesh of my youth felt, with its succession of military coups and its constant strikes, in the ensuing decades, politics in the country of my birth appears to have become even more contestable and contingent on its pastness. Despite a democratic transition in 1991 and the subsequent election of four parliamentary governments (albeit in hotly contested and sometimes controversial elections and with an interregnum of another military government between), the political debates of today are still marked by the memories and legacy of the 1971 war to an unusual extent and by the claims it continues to make on the present.

THE SHAHBAGH PROTESTS, DEMOCRATIC EMERGENCE, AND THE POLITICS OF BECOMING

In recent months, it has become abundantly clear the extent to which Bangladeshi politics is contingent on this unsettled past, which continually bursts through the present and pounds its fist on the table and demands to be heard. Earlier this year, Bangladesh saw a fascinating and somewhat spontaneous set of protests in the streets that seemed to signal an authentic moment of democratic emergence. Known popularly as the "Shahbagh" protests after the major traffic intersection they occupied in the nation's capital, the protestors drew inspiration from the democratic protests in Tahrir Square in Egypt in 2011 and the global Occupy movement. But while the Shahbagh was similar to these other movements in its democratic aspirations for greater freedom of expression and secularism in Bangladesh, its primary demand was of a rather different kind, one that made it difficult for international democratic activists to embrace wholeheartedly. That is because the key demand of the Shahbagh protests, vibrant as they were in their democratic becoming, clamored for the death penalty for a former *razakar* who they felt that the courts had sentenced far too leniently.

The Shahbagh protests appeared to signal a new form of political mobilization that was geared toward a different future but the horizon of that future was interwoven with the demands of the past. Most media accounts describe the protests as emerging from the grassroots and many describe them as spontaneously erupting after the Supreme Court sentenced Abdul Quader Mollah, a leader of the opposition Jamaat-E-Islami party, who was charged with war crimes during the Liberation War and of collaborating with the Pakistani Army, to life in prison. The life sentence was deemed far too lenient for the protestors who filled the busy Shahbagh intersection in the capital of Dhaka with insistent demands that he receive the death penalty instead. As with other recent global protests, social media played an important part in the organization of the Shahbagh movement, with a number of local bloggers playing a key role, and as the protests grew larger and more sustained over a number of days in February, the young activists kept up a large presence on both Facebook and Twitter to expand their reach and scope.[7] In a sense, one could argue that these massive popular protests were fueled by the quest for a kind of judicial revenge, but, from its very beginning, Shahbagh aimed to be something less coherent and more wide-ranging than that in its goals. Indeed many of the posts by protestors on social media signaled a deep frustration that the international media were not covering the protests more and giving them their due as the Tahrir of Bangladesh. The young activists wanted to be taken seriously in their demands for a new kind of politics and to be validated in this moment of democratic emergence but the protests themselves were interwoven with pastness—they remained wedded to the past, to a particular narrative of the Liberation War for which they sought justice and accountability, and the gestures towards the future that they offered were few and far between. As Jim Yardley noted in the *New York Times*,

> Huge daily demonstrations in the heart of Dhaka, the capital of Bangladesh, are upending the nation's politics and illustrating how heavily the country's bloody past still weighs on its present . . . Protests and strikes, common in Dhaka, are often coordinated and organized by political parties. But the Shahbagh protests, as the demonstrations over the verdict have come to be known, were organized by bloggers and have attracted poets, artists, social activists and untold numbers of other citizens.[8]

But from their inception the protests were about something more than a call for the death penalty, which, in many ways, was a proxy for a larger set of demands for justice and political accountability in a country that has gotten used to massive corruption in its politics and has little faith in the neutrality of its judicial system. As the *New York Times* reported, Sultana Kamal, a prominent human rights activist, said that she "disagreed with the calls for the death penalty, but that they reflected the cynicism of Bangladeshis who have seen war criminals evade punishment for decades. Many people were infuriated when Mr. Mollah flashed a victory sign after receiving his life sentence. "We have a problem in accepting that they are demanding the death penalty", Ms. Kamal said in a telephone interview. "But we understand that it was from a nervousness among the people here that unless they are given the highest penalty in the land, these people will come back out".[9] As such, while the protestors' demand for the death penalty could be seen as a gory call for revenge for the war crimes of the 1971 war, their very mobilization in the present, across segments of the population have rarely participated in politics before is nonetheless significant. As a senior fellow at the Center for Policy Dialogue, a leading research institute in Dhaka, noted, "This is something different and something new . . . This is the rise of a new social force that can change the political calculus in the country".[10]

But the possibility signaled by the emergence of this "new social force" was also imbricated in a complex politics of becoming. The newness of Shabagh still operated within a contingent and layered network of forces of history, power, and aspiration. While the euphoria of the crowds in Shahbagh suggested something new was afoot, they were still treading on the shaky ground of a troubled past that remained unsettlingly present. As mentioned earlier, little has been done until recently to investigate the war crimes of the 1971 war or to account for that past. In 2011 the Awami League led government embarked on an effort to put several alleged *razakars* on trial, before what they called the International War Crimes Tribunal, though the courts themselves have no connection to the International Criminal Court, nor do they operate according to internationally recognized rules of law. As such, while a set of war crimes trials are taking place and guilty verdicts are regularly being delivered, the absence of investigation and documentation, the lack of due process, and the politicized nature of the trials (all of the accused, though widely known to be collaborators with the former Pakistani army, are now also leaders of the Jamaat-E-Islami political party and play a crucial role in the opposition alliance) make accountability difficult to deliver. An op-ed in the *New York Times* described some of the objections to the tribunal, noting that rather than helping to stabilize the injured past, it

> serves as a terrible warning of the way in which the ideals of universal justice and accountability can be abused . . . To make a considered moral judgment on a conflict that took place 40 years ago, a scrupulously impartial investigation would be needed. Sadly, the current trial promises to be nothing of the kind. It pretends to be applying universal principles—that is implicit in the name of the court, the Bangladesh International Crimes Tribunal—but in contrast with other recent ad-hoc tribunals, there is no external input, because none has been allowed.[11]

As the trials unfolded, more evidence of flaws and wrongdoing in the judicial process emerged. In addition to lack of proper procedure such as access to lawyers by the defendants, there were accounts of prolonged detention and interrogation, and perhaps most disturbingly, Skype conversations between one of the justices and a Brussels-based lawyer, in which they discussed intense political pressure by the government to deliver a particular verdict, were leaked to the media.[12]

The flawed nature of the trials meant that, first, there was not much faith that the tribunal could create accountability for the past and, second, that for many young Bangladeshis, the seeming failure of the court to deliver what they considered to be an appropriate verdict meant that political solutions to the presence of the past had to be sought elsewhere. As one observer noted, "Young kids baying for blood will make many justifiably uncomfortable. And no court should be forced to alter a verdict because of popular pressure. But the protests in Shahbagh must be seen as more than a demand for blanket death sentences . . . The protests represent the spirit of 1971—and a bigger war that is yet to be won".[13]

Despite its glances at a democratic future, and even as it jolted and destabilized the existing political structures, which scrambled to co-opt and make sense of them, the Shahbagh protests came into being precisely because of the paralyzing hold of the past. In a sense they did represent "the spirit of 1971", a ghostly spirit that remained disquietingly present. Even as the protests themselves gestured at a moment of political becoming and offered a brief glimpse of democratic emergence, they served to further underline the extent to which Bangladeshi politics was contingent on its unsettled past.

Politics is always contingent on moments of becoming, on alliances that shift beneath our feet, but it also permeated by a restless anxiety about a future that is not quite here that remains harnessed to a past that never quite lets go. In the case of the Shahbagh protests, that past is a particularly disconcerting and violent one that has not and perhaps cannot be adequately accounted for within the existing governing structure. In many ways Shahbagh represented a search for something new, a moment of becoming that at once strained toward a different future but remained contingent on the forces of history that also propelled it into that uncertain horizon. The Shahbagh movement brings to mind one of Walter Benjamin's reflections from the *Theses on the Philosophy of History*, where he writes, To articulate the past does not mean to recognize it 'the way it really was' (Ranke). It means to seize hold of a memory as it flashes up at a moment of danger . . . In every era the attempt must be made anew to wrest tradition away from a conformism that is about to overpower it".[14] Perhaps Shahbagh represents just that effort that Benjamin described—the attempt to seize hold of a memory. The movement arose as the memory of 1971 flooded back into the present with the too-lenient verdict of Mollah, a moment that perhaps flashed up dangerously—not because of the potential for further political destabilization but because it revealed the depths and enormity of the visceral hold that the 1971 war continued to have on the nation's collective present. Accountability for the war was an urgent demand of the protests, but equally significant was the newness of their form and the extent to which they served as a critique of the existing political structure. In that, Shahbagh may well have wrenched tradition away from the overpowering conformism of everyday corruption and politicized justice. But even that attempt remained contingent on the flux and flow of election year politics and by others who also laid claim to that past. The potential of Shahbagh seemed to fade almost as quickly as it emerged, followed as it was by months of protests by Islamists and the violent suppression of those protests by the government. Even before Shahbagh could coalesce into a movement that could mobilize for real and sustained political change, it was overtaken by the same paralyzing forces of the past that seemingly refuse to lessen their grip on the present. The Shahbagh protests were indeed a form of democratic emergence and a fugitive moment of political becoming, but as much as they signified the new, they also served to highlight how any political orienting towards the future is also always a reorientation in a particular past. That attempt to wrest tradition away from conformism has the potential to open up the future in unsettling and possibly liberating ways but it also underlines the notion that the Shahbagh protests existed and perhaps continue to exist in a time of the political that remains out of joint.

THE POLITICS OF CONTINGENCY

Operating in a world where politics is nudged and jostled by a ghostly past, by the veritable presence of the Shakespearean wraith, is to underline the extent to which contingency is linked to the understanding of temporality.

To inhabit a political that is out of joint is to acknowledge its tragic quality, but it is also to appreciate its politics of becoming, to attend to the unsettled layers of sedimentation that shift and move and occasionally crack, but on which we build our lives and our sources of meaning. Memory, which we fall into, but which also floods us in the moment, is the mark of contingency. We cannot help but play the game, to cast Nietzsche's famous dice throw, which "affirms becoming" and "the being of becoming".[15] In the past year, Shahbagh was one such dice throw, one attempt at both chance and necessity, an attempt both startlingly new and somehow inevitable. It is this combination of forces, of the being of becoming, of the necessity of chance, of the recognition of the past that makes action in the present so challenging that undergirds the politics of contingency.

Pastness and its continued existence in the present make political action difficult. It can result in the paralysis of inaction. It can strangle a moment of democratic becoming before it ever coalesces into something resembling a movement. It is this confluence between the recognition of time as out of joint and the tragic understanding of the necessity of affirming and playing the game that is a mark of the politics of contingency. William Connolly brings this set of relations to the forefront when he writes that "we belong to time, but we do not often think about the strange element through (or 'in') which we live, breathe, act, suffer, love, commune, and agitate. Indeed, it would be unwise if we focused on this register of experience too often. We would lose our ability to act with efficacy, confidence, and fervor in the world. For action requires simplified perception to inform it".[16] While Connolly highlights the relationship between recognizing time as out of joint and political action, he also suggests that it is "unwise" to focus on this too much. In part, this is because Connolly's focus here is unapologetically on the ways in which positive memories can help our politics and institutions by mobilizing actions and ethical sensibilities through micro politics. As such, he deals with the question of trauma and the paralysis it can engender in just a couple of sentences. To the limited extent that I disagree with Connolly here, it is with his unfiltered optimistic assessment for the possibility of political action because I argue that the political reality is far more tragic in the sense that Nietzsche describes. Attending to the notion that we belong to time, or understanding its richness, can help one grasp the essential contingency of politics, as existing between a multitude of agents, each with their many pasts and spaces of experience and horizons of futures, a rhizome of sometimes muted, sometimes forgotten, sometimes recollected networks of temporal multiplicity. And while such recognition can deepen and enrich our action and perception of politics by drawing our attention to moments of emergence, inaction by the paralysis of the past remains a pressing and forbidding threat. As seen in the case of the Shahbagh protests in Bangladesh, the past also threatens to envelop, to refuse legislative and ethical attempts at justice on the individual and collective level, and choke off moments of democratize emergence before they can crystallize.

What Connolly calls "tragic possibility"[17] (even if his account of that possibility is less tragic than it could have been) is what the politics of contingency must attend to, the knowledge of both the transformative power of newness that can explode the temporal spectrum in a messianic moment of change that Benjamin so famously described, or the moment of paralysis, where even the roiling forces of agency are frozen in place by a past that does not give way.[18] Not only does such a recognition underscore that do we belong to time, but the recognition of time as out of joint also is a fundamental recognition of the politics of contingency. It means that the tragic dice throw we must all cast is replete with possibility but also ordered by chance. This sense of a world that is no longer structured by faith in the promise of salvation that Augustine once described, nor regulated by the laws of reason that Kant outlined, leads to an understanding of the political as necessarily contingent on disruption and disquiet.

What does such an understanding mean for political action? One such possibility is the one that Connolly notes: that "the operation of multiple tiers of becoming in a world without a higher purpose amplifies the need to act with dispatch, and sometimes with militancy, in particular situations of stress".[19] But it also sometimes leads to inaction, to the inability to effect change as in the case of Shahbagh, to voices that are silenced by those in positions of power that do not operate with the same temporal recognition as in the case of the Afghani prisoner Wali, whose story we encountered in the first chapter, whose experience would always remain incommensurable with that of his interrogators; to the victims of civil strife and genocide in Algeria, South Africa, and Rwanda discussed in Chapter 4, who struggled to find collective solutions to the stress of the past; and to those who attempt to negotiate and make sense of a collapsing binary between Islam and the West that rests on a particular culturally constituted temporal foundation that is dented and cracked by the recognition of multiplicity that was explored in Chapter 5. Indeed, as Connolly says, "the quest for eternal salvation and the possibility of human mastery over nature are both challenged by a world of becoming".[20] But what kind of politics is challenged by the recognition of time as out of joint? An understanding of the political that is centered solely on the present, which assumes that the past no longer plays a constitutive and disciplining role in the present, is certainly challenged by such an account. To be clear, it is not so much that all analysis of politics or forms of political action need to be steeped in historicity, but rather that to be aware of the roiling subterranean forces that impact the present allow for a greater understanding of both contestation and difference in the political.

Becoming attentive to the inherent contingency of the political can make us more aware not only of the complexity and fragility of emergence but also of the real possibility of the new and its emergent capacity to jostle ossified formations. This sense of the tragic possibility that is at the core of coming to terms with the politics of contingency is captured lyrically by Nietzsche in *Thus Spoke Zarathustra*. He writes, "Still is the bottom of my sea, who would guess it harbors sportive monsters? Imperturbable is my depth, but it sparkles with swimming riddles and laughters".[21] Nietzsche's evocative quote suggests a true field of immanence in which seemingly stable states are always populated by mischievous and unpredictable flows of various intensities that are forever changing and forever becoming. It implies not a rigid stratum in which the various registers of being are sealed off separately from each other or united by a simple notion of linear progress but instead a rhizomatic assemblage of various forces that flow in and out of each other and hence engage in a truly non-linear relationship. In *A Thousand Plateaus*, Gilles Deleuze and Felix Guattari suggest that the sparkling depths and the swimming riddles that Nietzsche invokes are crucial to and form the basis of a true philosophy of immanence, which does not anchor itself in the imperatives and the limits of transcendence but seeks to explore the flows of becoming. As such, Deleuze and Guattari "do not *reflect* upon the world but are *immersed* in the changing order of things" (emphasis added).[22] Theirs is a philosophy of the rhizome, of nomad thought that revels in ruptures of sequence and in the possibility of splintering time, a philosophy of flows and meshworks of varying intensities that relies profoundly on an understanding of contingent networks of forces for its very creation and its sufficiency for an ethical life.

The very invocation of contingency suggested by Nietzsche's quote brings with it the attendant concept of time as out of joint, as unpredictable and full of rifts and forks that are pregnant with an explosive creativity. Such a recognition of the rupture and possibility of the politics of contingency underlines the insufficiency of both the Kantian laws of universal morality and the structuring demands of reason and religion that seek to impose order on this forever changing world; they also call attention to the quieting impulse of the disciplinary regimes of power that erase multiplicity and cultural constitution in the name of universalism. An understanding of the politics of contingency holds out hope for an ethics that is receptive to difference and embraces the possibility of joy that can allow us to engage in a more vigorous pluralism. But it also draws attention to the paralysis of the past and its spectral hold on the present and to how that past might actively be transfigured in the present for the sake of the future, or how it can continue to unsettle and strangle new forms of political action. In the end, what we might to be able to draw out of an exploration of the politics of contingency is an attentiveness to the multiple and vigorous forces that seem to occur beneath the seemingly stable coagulations of the political, to a recognition of the sportive monsters that inhabit our depths. A philosophical and ethical orientation to these networks of forces may not result in the dissolution of unity but instead in recognition of the complexity of the formations political and ethical life. And as we attend to the continued demands of the out of joint experience of time, we can not only perhaps acknowledge that the past sometimes brings with it grief, trauma, and an inability to act in the present, but we can also hold out hope for the possibility of joy amidst such recognition, look below the roiling depths of the present, and entertain the possibilities of swimming riddles and laughter.

NOTES

1. Sarmila Bose, *Dead Reckoning: Memories of the 1971 Bangladesh War* (New York: Columbia University, Press, 2011), p. 20.
2. Rape, and sexual violence in particular, is rarely referred to in the omnipresent discourse about the war. The 300,000 number is part of the familiar refrain of the staggering costs of the war, but little attention has been given to interviewing victims or exploring the consequences of this immense violence. In an interview with the *New York Times*, Irene Khan, former secretary-general of Amnesty International, put it this way: "A conservative Muslim society has preferred to throw a veil of negligence and denial on the issue, allowed those who committed or colluded with gender violence to thrive, and left the women victims to struggle in anonymity and shame and without much state or community support"; Nilanjana S. Roy, "Bangladesh's War Toll on Women Still Undiscussed", *The New York Times*, August 24, 2010.
3. Syed Badrul Ahsan, "Ground Realities, Sarmila Bose and Bad Arithmetic", op-ed, *The Daily Star*, July 13, 2011.
4. Martin Woollacott, "Dead Reckoning by Sarmila Bose—Review A Long-Overdue Study of Bangladesh's War of Independence", *The Guardian*, July 1, 2011.
5. Amber Abbas, "Review of Bose, Sarmila, *Dead Reckoning: Memories of the 1971 Bangladesh War*", *H-Memory, H-Net Reviews*, March, 2012, www.h-net.org/reviews/showrev.php?id=34415.
6. Bose, *Dead Reckoning*, p.183.
7. For a representative collection of social media activism and posts and video of the protests see "Bangladesh's Rising Voices: War Crimes Protests Heat Up across the Nation", in *The Stream* (Al-Jazeera), http://stream.aljazeera.com/story/201302190008–0022555.
8. Jim Yardley, "Politics in Bangladesh Jolted by Daily Demonstrations", *The New York Times*, February 12, 2013.
9. Julfikar Ali Manik and Jim Yardley, "Death Toll From Bangladesh Unrest Reaches 44", *The New York Times*, March 1, 2013.
10. Yardley, "Politics in Bangladesh."
11. John Cammegh, "In Bangladesh: Reconciliation or Revenge", op-ed, *The New York Times*, November 17, 2011.
12. Sabir Mustafa, "Bangladesh's Watershed War Crimes Moment", *BBC News Asia*, January, 21, 2013, www.bbc.co.uk/news/world-asia-21133320.
13. Shahidul Islam, "A 40-Year Quest for Justice", op-ed, *The New York Times*, February 28, 2013.
14. Walter Benjamin, "Theses on the Philsophy of History", in *Illuminations*, ed. Hannah Arendt and trans. Harry Zohn (New York: Schocken Books, 1968), p. 255.
15. Gilles Deleuze, *Nietzsche & Philosophy*, trans. Hugh Tomlinson (New York: Columbia University Press, 1983), p. 25.
16. William E. Connolly, *A World of Becoming* (Durham, NC: Duke University Press, 2011), p. 2.
17. Ibid., p. 7.
18. A useful account of nonlinear time can also be found in Walter Benjamin's "Theses on the Philosophy of History." Benjamin theorizes a "now time",

which consists of a present that is not merely a transition in the quantified continuum between the past and the future, but a present in which time stands still. It is an explosive moment that creates an energized and distinctively new "now-time" of an innovative present that interrupts the continuum and breaks its sluggish flow. See the discussion of Messianic Time in Benjamin, "Theses on the Philosophy of History", pp. 253–264.

19. Connolly, *A World of Becoming*, p. 7.
20. Ibid., p. 10.
21. Friedrich Nietzsche, *Thus Spoke Zarathustra*, trans. Walter Kaufmann (New York: Penguin Books, 1978), p. 116.
22. Brian Massumi, "Translator's Foreword", in *A Thousand Plateaus: Capitalism and Schizophrenia* by Gilles Deleuze and Felix Guattari (Minneapolis: University of Minnesota Press, 1987), p. xii.

Bibliography

Abbas, Amber. "Review of Bose, Sarmila, *Dead Reckoning: Memories of the 1971 Bangladesh War*." *H-Memory, H-Net Reviews*, March, 2012. www.h-net.org/reviews/showrev.php?id=34415.

Agamben, Georgio. *Infancy and History: Essays on the Destruction of Experience*. Translated by Liz Heron (New York: Verso Books, 1993).

Ahsan, Syed Badrul. "Ground Realities, Sarmila Bose and Bad Arithmetic", op-ed. *The Daily Star*, July 13, 2011.

"Aims and Objectives of the Charter". *Algeria Today*. September 30, 2005. www.algeriaus.org/ALGERIA_TODAY_0%5b1%5d.pdf.

Akbar, Hyder. "*Memoir: Interrogation Unbound*. As Told to Susan Burton". *The New York Times Magazine*, section 6, July 11, 2004.

"Algeria Referendum Numbers Queried". *Africa Research Bulletin: Political Social and Cultural Series* 42, no. 10 (October 1–31, 2005): 16383–16384.

Anderson, Benedict. *Imagined Communities: Reflections on the Origin and Spread of Nationalism* (London: Verso, 1991).

Asad, Talal. *Formations of the Secular: Christianity, Islam, Modernity* (Stanford, CA: Stanford University Press, 2003).

Augustine. *City of God*. Translated by Henry Bettenson (London: Penguin Books, 1972).

———. *The Confessions*. Translated by M. Boulding (New York: Vintage Spiritual Classics, 1997).

"Bangladesh's Rising Voices: War Crimes Protests Heat Up across the Nation". In *The Stream* (Al-Jazeera). http://stream.aljazeera.com/story/201302190008–0022555.

Benjamin, Walter. *Illuminations*. Edited by Hannah Arendt and translated by Harry Zohn (New York: Schocken Books, 1968).

Bergson, Henri. *Matter and Memory*. Translated by N. M. Paul and W.S. Palmer (New York: Zone Books, 2002).

Blumenberg, Hans. *The Legitimacy of the Modern Age*. Translated by Robert E. Wallace (Cambridge, MA: MIT Press, 1983).

Bose, Sarmila. *Dead Reckoning: Memories of the 1971 Bangladesh War* (New York: Columbia University, Press, 2011).

Brown, Wendy. *Politics Out of History* (Princeton, NJ: Princeton University Press, 2001).

Burns, Matthew. "Sentence Reduced for CIA Contractor Convicted of Detainee Abuse". *New York Times*, April 6, 2010. www.wral.com/news/news_briefs/story/7366849/.

Butler, Judith. *Frames of War: Precarious Life, Grievable Life* (New York: Verso, 2009).

———. "Sexual Politics, Torture, and Secular Time". *The British Journal of Sociology* 59, no. 1 (2008): 1–23.

Cammegh, John. "In Bangladesh: Reconciliation or Revenge", op-ed. *The New York Times*, November 17, 2011.
Carvounas, David. *Diverging Time: The Politics of Modernity in Kant, Hegel and Marx* (Oxford: Lexington Books, 2002).
Caygill, Howard. *A Kant Dictionary* (Cambridge, MA: Blackwell Publishers, 1995).
Chambers, Samuel A. *Untimely Politics (Taking on the Political)* (New York: New York University Press, 2003).
———. "Untimely Politics Avant la Lettre: The Temporality of Social Formations". *Time & Society* 20 (July 20): 197–223.
Clark, Philip. *The Gacaca Courts, Post-Genocide Justice and Reconciliation in Rwanda: Justice without Lawyers* (Cambridge: Cambridge University Press, 2010).
———. "The Legacy of Rwanda's Gacaca Courts". *Think Africa Press*, March 23, 2012. http://thinkafricapress.com/rwanda/legacy-gacaca-courts-genocide.
———. "When the Killers Go Home: Local Justice in Rwanda". *Dissent Magazine* (Summer 2005): 14–21.
Connolly, William E. *Neuropolitics: Thinking, Culture, Speed* (Minneapolis: University of Minnesota Press, 2002).
———. *Why I Am not a Secularist* (Minneapolis: University of Minnesota Press, 2000).
———. *A World of Becoming* (Durham, NC: Duke University Press, 2011).
Deleuze, Gilles. *Bergsonism*. Translated by Hugh Tomlinson and Barbara Habberjam (New York: Zone Books, 1991).
———. *Cinema 2: The Time-Image*. Translated by Hugh Tomlinson and Robert Galeta (Minneapolis, MN: University of Minnesota Press, 1997).
———. *Difference & Repetition*. Translated by Paul Patton (New York: Columbia University Press, 1994).
———. *Nietzsche & Philosophy*. Translated by Hugh Tomlinson (New York: Columbia University Press, 1983).
———. *A Thousand Plateaus: Capitalism and Schizophrenia*. Translated by Brian Massumi (Minneapolis: University of Minnesota Press, 1987).
Derrida, Jacques. *Specters of Marx: The State of the Debt, the Work of Mourning and the new International*. Translated by Peggy Kamuf with an introduction by Bernd Magnus and Stephen Cullenberg (New York and London: Routledge, 1994).
Dienstag, Joshua Foa. "Tragedy, Pessimism, Nietzsche". *New Literary History* 35, no. 1 (Winter 2004): 83–101.
Duvenage, Pieter. "The Politics of Memory and Forgetting after Auschwitz and Apartheid". *Philosophy and Social Criticism* 25, no. 3 (1999): 1–28.
Euben, Roxanne L. *Enemy in the Mirror: Islamic Fundamentalism and the Limits of Modern Rationalism* (Princeton, NJ: Princeton University Press, 1999).
Faulkner, William. *Requiem for a Nun* (New York: Vintage Books, 1975).
"A Flawed Charter". *The Economist*, 337, no. 8447 (October 8, 2005): 56.
Foucault, Michel. *The Foucault Reader*. Edited by Paul Rabinow (New York: Random House, 1984).
Gibson, James L. "Overcoming Apartheid: Can Truth Reconcile a Divided Nation?" *Politikon: South African Journal of Political Studies* 31 (November 2004): 129–155.
Grosz, Elizabeth. *The Nick of Time: Politics, Evolution and the Untimely* (Durham, NC, and London: Duke University Press, 2004).
Habermas, Jürgen. *Habermas and the Unfinished Project of Modernity: Critical Essays on the Philosophical Discourse of Modernity*. Edited by M. Passerin d'Entrèves and Seyla Benhabib (Cambridge, MA: MIT Press, 1996).

———. *The Philosophical Discourse of Modernity.* Translated by Frederick G. Lawrence (Cambridge, MA: MIT Press, 1995).

Heidegger, Martin. *Nietzsche. Vol. II: The Eternal Recurrence of the Same.* Translated by D.F. Krell (San Francisco: Harpers & Row, 1979).

Ignatieff, Michael. "The Elusive Goal of War Trials". *Harper's Magazine* 294, no. 1762 (March 1997): 15–19.

Isaac, Jeffrey. "The Strange Silence of Political Theory". *Political Theory* 23, no. 4 (November 1995): 636–652.

Islam, Shahidul. "A 40-Year Quest for Justice", op-ed. *The New York Times*, February 28, 2013.

Joffé, George. "National Reconciliation and General Amnesty in Algeria". *Mediterranean Politics* 13, no. 2 (July 2008): 213–228.

Kant, Immanuel, *Political Writings.* Edited by H. Reiss and translated by H. B. Nisbet (Cambridge: Cambridge University Press, 1991).

———. *Religion within the Limits of Reason* (New York: Harper & Row Publishers, 1990).

Kaufmann, Walter, *Nietzsche: Philosopher, Psychologist, Antichrist* (Princeton, NJ: Princeton University Press, 1974).

Koselleck, Reinhart. *Futures Past: On the Semantics of Historical Time* (New York: Columbia University Press, 2004).

Mahmood, Saba. "Secularism, Hermeneutics and Empire: The Politics of Islamic Reformation". *Public Culture* 18, no. 2 (2006): 323–347.

Manik, Julfikar Ali, and Yardley, Jim. "Death Toll From Bangladesh Unrest Reaches 44". *The New York Times*, March 1, 2013.

Markell, Patchen. *Bound by Recognition* (Princeton, NJ: Princeton University Press, 2003).

Massumi, Brian. "Translator's Foreword", in *A Thousand Plateaus: Capitalism and Schizophrenia* by Gilles Deleuze and Felix Guattari (Minneapolis: University of Minnesota Press, 1987).

McEwan, Cheryl. "Building a Postcolonial Archive? Gender, Collective Memory and Citizenship in Post-Apartheid South Africa". *Journal of Southern African Studies* 29, no. 3 (September 2003): 739–757.

Michalson, Gordon E. *Fallen Freedom* (Cambridge: Cambridge University Press, 1990).

Minow, Martha. *Between Vengeance and Forgiveness: Facing History after Genocide and Mass Violence* (Boston: Beacon Press, 1998).

Mustafa, Sabir. "Bangladesh's Watershed War Crimes Moment". *BBC News Asia*, January 21, 2013. www.bbc.co.uk/news/world-asia-21133320.

National Unity and Reconcilation Commission (NURC). *Social Cohesion in Rwanda, An Opinion Survey, Results 2005–2007.* www.nurc.gov.rw/index.php?id=69&eID=dam_frontend_push&docID=12.

Nietzsche, Friedrich. *The Birth of Tragedy.* In *The Birth of Tragedy and the Case of Wagner,* translated by Walter Kaufmann (Toronto: Vintage Books, 1967).

———. *Thus Spoke Zarathustra.* Translated by W. Kaufmann (New York: Penguin Books, 1978).

———. *Untimely Meditations,* Trans, R.J. Hollingdale. (Cambridge: Cambridge University Press, 1997).

———. *The Will to Power.* Edited by Walter Kaufmann and translated by Walter Kaufmann and R. J. Hollingdale (New York: Vintage Books, 1968).

Pocock, J.G.A. *The Machiavellian Moment: Florentine Political Thought and the Atlantic Republican Tradition* (Princeton, NJ: Princeton University Press, 1975).

Power, Samantha. *A Problem from Hell: America and the Age of Genocide* (New York: Perennial Books, 2003).

The Promotion of National Unity and Reconciliation Act (95-34, July 26, 1995). www.doj.gov.za/trc/legal/act9534.htm.
Qutb, Sayyid. *Milestones* (Damascus: Kazi Books, 1964).
Roy, Nilanjana S. "Bangladesh's War Toll on Women Still Undiscussed". *The New York Times*, August 24, 2010.
Shane, Scott, "C.I.A. Contractor Guilty in Beating of Afghan Who Later Died". *The New York Times*, August 18, 2006. http://select.nytimes.com/gst/abstract.html?res=F20910FE3E5A0C7B8DDDA10894DE404482&emc=eta1.
Shakespeare, William. *The Tragedy of Hamlet, Prince of Denmark*. In *The Oxford Shakespeare: The Complete Works, Second Edition*, edited by Stanley Wells and Gary Taylor, 681–718 (Oxford: Oxford University Press, 2005).
Slackman, Michael. "Top Algerians Prefer Amnesia to Accountability on War". *The New York Times*, September 26, 2005
———. "But Bygones Can't Be Bygones if the Pain Is Raw". *The New York Times*, October 5. 2005.
Spiegel, Alix. "Can We Learn to Forget Our Memories?" *National Public Radio*, September 3, 2012. www.npr.org/blogs/health/2012/09/03/160126283/can-we-learn-to-forget-our-memories.
Strain, Jason, and Keyes, Elizabeth. "Accountability for Atrocities in the Aftermath of Rwanda's Genocide". In *Accountability for Atrocities: National and International Responses*, edited by Jane E. Stromseth, 87–133 (Ardsley, NY: Transnational Publishers, Inc, 2003).
Taylor, Charles. "Modes of Secularism". In *Secularism and its Critics*, edited by Rajeev Bhargava, 31–53 (Delhi: Oxford University Press, 2005).
———. *A Secular Age* (Cambridge, MA: Belknap/Harvard University Press 2007).
Trouillot, Michel-Rolph. *Silencing the Past: Power and the Production of History* (Boston: Beacon Press, 1995).
Wolin, Sheldon. "What Time Is It?" *Theory and Event* 1, no. 1 (1997). http://muse.jhu.edu/journals/tae.
Woollacott, Martin. "Dead Reckoning by Sarmila Bose—Review A Long-Overdue Study of Bangladesh's War of Independence". *The Guardian*, July 1, 2011.
Yardley, Jim. "Politics in Bangladesh Jolted by Daily Demonstrations". *The New York Times*, February 12, 2013.

Index

For Product Safety Concerns and Information please contact our EU representative GPSR@taylorandfrancis.com
Taylor & Francis Verlag GmbH, Kaufingerstraße 24, 80331 München, Germany

www.ingramcontent.com/pod-product-compliance
Ingram Content Group UK Ltd.
Pitfield, Milton Keynes, MK11 3LW, UK
UKHW021958180926
475701UK00003B/80
9781138687011